PHYSICS RESEARCH AND TECHNOLOGY

OPTOELECTRONICS IN MEASUREMENT OF PHYSICAL MAGNITUDES

PHYSICS RESEARCH AND TECHNOLOGY

Additional books in this series can be found on Nova's website under the Series tab.

Additional E-books in this series can be found on Nova's website under the E-book tab.

PHYSICS RESEARCH AND TECHNOLOGY

OPTOELECTRONICS IN MEASUREMENT OF PHYSICAL MAGNITUDES

V. N. FEDORININ
A.G. PAULISH
AND
A. S. LEVINA

Nova Science Publishers, Inc.
New York

Copyright © 2011 by Nova Science Publishers, Inc.

All rights reserved. No part of this book may be reproduced, stored in a retrieval system or transmitted in any form or by any means: electronic, electrostatic, magnetic, tape, mechanical photocopying, recording or otherwise without the written permission of the Publisher.

For permission to use material from this book please contact us:
Telephone 631-231-7269; Fax 631-231-8175
Web Site: http://www.novapublishers.com

NOTICE TO THE READER

The Publisher has taken reasonable care in the preparation of this book, but makes no expressed or implied warranty of any kind and assumes no responsibility for any errors or omissions. No liability is assumed for incidental or consequential damages in connection with or arising out of information contained in this book. The Publisher shall not be liable for any special, consequential, or exemplary damages resulting, in whole or in part, from the readers' use of, or reliance upon, this material. Any parts of this book based on government reports are so indicated and copyright is claimed for those parts to the extent applicable to compilations of such works.

Independent verification should be sought for any data, advice or recommendations contained in this book. In addition, no responsibility is assumed by the publisher for any injury and/or damage to persons or property arising from any methods, products, instructions, ideas or otherwise contained in this publication.

This publication is designed to provide accurate and authoritative information with regard to the subject matter covered herein. It is sold with the clear understanding that the Publisher is not engaged in rendering legal or any other professional services. If legal or any other expert assistance is required, the services of a competent person should be sought. FROM A DECLARATION OF PARTICIPANTS JOINTLY ADOPTED BY A COMMITTEE OF THE AMERICAN BAR ASSOCIATION AND A COMMITTEE OF PUBLISHERS.

Additional color graphics may be available in the e-book version of this book.

LIBRARY OF CONGRESS CATALOGING-IN-PUBLICATION DATA

Fedorinin, V. N. (Viktor Nikolaevich)
  [Poliarizatsionno-opticheskie pribory dlia izmereniia fizicheskikh velichin. English]
  Optoelectronics in measurement of physical magnitudes / V.N. Fedorinin and A.G. Paulish, A.S. Levina.
    p. cm.
  Includes index.
  ISBN 978-1-61761-094-3 (softcover)
  1. Optoelectronics. 2. Polarization (Light) 3. Electromagnetic measurements. I. Paulish, A. G. (Andrei Georgievich) II. Levina, A. S. (Asia Saulovna) III. Title.
  TA1750.F4313 2010
  621.381'045--dc22
                                      2010028559

Published by Nova Science Publishers, Inc. † New York

CONTENTS

Preface		vii
Introduction		1
Chapter 1	The Polarization-Optical Detector	3
Chapter 2	Measurement of Infra-Red Radiation Intensity	7
Chapter 3	Measurement of the Acoustic Signal Spectrum	17
Chapter 4	Measurement of the Deformation	27
Chapter 5	Measurement of the Physical and Chemical Parameters of Liquid and Gas	33
Chapter 6	The DNA Sensor	45
Chapter 7	Conclusion	57
References		59
Index		63

PREFACE

The development of optoelectronic devices for the measurement of physical magnitudes, such as electromagnetic radiation intensity with spatial resolution, acoustic signal spectrum, magnitudes of the displacement, and kinetics of chemical reactions are presented. The operation principle of given devices is based on the change in a state of electromagnetic wave polarization upon the interaction with physical object while changing its mechanical and chemical characteristics.

The infra-red radiation detector with spatial resolution is based on the matrix structure of optoacoustic cells (so-called Gollay cells). Its receiving element is a radiation absorbing film which heats gas in a small enclosed chamber. A flexible wall (membrane) of the chamber distorts under the pressure of the heated gas, and this gives a measure of absorbed power. High responsivity is achieved with an optical measurement of wall deflection determined from the deflection of a light beam reflected from distorted membrane and detected by a photodetector. The optical readout system developed in this contribution includes the Savart plate, visible light-emitting diode (LED), commercial CMOS camera and has the sensitivity to the membrane displacement below 1 nm. The uncooled matrix detector of infra-red radiation is designed on the basis of the proposed system.

The optical readout system was used for the acoustic signal analysis. Microstrings of various length and diameter with resonant frequencies over the range of 100–20000 Hz are used as the resonant sensors of the acoustic vibrations. The design of the optical readout system based on the CMOS camera and procedure of the acoustic signal frequency spectrum definition are proposed.

The piezo-optical (photoelastic) effect was used for the development of high-sensitive linear strain gauges.

The property of electromagnetic radiation to change the polarization at the reflection from interfaces was used for the development of gauges of physical and chemical parameters of liquid, gas, and biological environments. The portable optoelectronic refractometer and the gas sensor are presented. The DNA sensor based on the polarization detector was developed for the analysis of viral or bacterial nucleic acids responsible for dangerous diseases.

Main advantages of all these devices are the use of the noncoherent light source (LED) and the absence of a necessity to develop special electronic readout systems because wires are not required for the light transfer. As a result, such devices are compact, inexpensive, and are constructed on the unified principle.

INTRODUCTION

The basic tendency of modern electronics is the continuous complication of the circuits at the increase of the informativity, technical and economic parameters, such as reliability, speed and the decrease of size and weight of devices. Thereupon, optoelectronics finds more and more application for information processing, storage and transmission mainly due to the rapid development of solid-state lasers, semiconductor light emitting diodes, and fiber optics.

Optoelectronics differs from vacuum and semiconductor electronics by the presence of optical communication in a signal circuit. Advantages of optoelectronics are defined, first of all, by the preference of optical communication over electrical one. Unlike the electric current-passing, there are no electrical and magnetic fields excited in the optical communication channel because of photons electrical neutrality. In other words, photons do not create the additional noise in communication lines and provide a full electrical separation between the transmitter and the receiver. That is essentially unattainable in circuits with electrical communication. The information transfer is not accompanied by the electromagnetic energy accumulation and dissipation. Therefore, there is no considerable signal lag in the communication channel that provides a high speed and a minimal sending-end distortion of the transferred information, which is transported by the optical signal.

The next advantages of optoelectronics are the opportunities resulting from the use of various physical phenomena caused by interaction of light fields with a solid state. That provides the development of the noncontact methods for the interface investigations. The most sensitive parameter to the light-solids interaction is the electromagnetic wave polarization. In this contribution, the series of the optoelectronic devices for the measurement of physical magnitudes, such as electromagnetic radiation intensity with spatial resolution, acoustic signal spectrum, magnitudes of the displacement, refractive index, and kinetics of chemical reactions are developed. The operation principle of proposed devices is based on the changes in the polarization state of the electromagnetic wave upon the light interaction with a physical object while changing its mechanical and chemical characteristics. These magnitudes are chosen because they are in the basis of human sense organs: vision, hearing, touch, smell, and taste. The development of analogous sensors is the important problem in the creation of an artificial intellect.

Chapter 1

THE POLARIZATION-OPTICAL DETECTOR

As is well known, the electromagnetic wave can be described by three mutually orthogonal vectors. The first one determines the direction of the electromagnetic energy flow; the next two describe the direction of the electric field intensity of the electromagnetic wave (Figure 1.1, top). In other words, the last two vectors describe the state of the electromagnetic wave polarization. Depending on the phase shift Δ between the two orthogonal vectors of the electric field intensity, the resultant vector generally circumscribes the ellipse. In particular cases, when $\Delta = 0°$ or $\Delta = 180°$, the ellipse degenerates in to the straight line (Figure 1.1, below), and this wave is called linearly polarized light. In cases $\Delta = 90°$ or $\Delta = 270°$ the resultant vector circumscribes the circle, and this wave is called circularly polarized light.

The changes in the state of the polarization of the electromagnetic wave upon the light interaction with a physical object while changing its mechanical and chemical characteristics, underlies the polarization-optical method and the detector developed in this work [1].

The detector consists of the source of light, polarizer, phase plate (quarter-wave plate), investigated sample, two photodetectors, and two analyzers established in front of the photodetectors (Figure 1.2 and Figure 1.3). The analyzers axes are turned to each other by $90°$. The light reflected from (or transmitted through) the sample is a space divided into two rays of equal intensity and detected by two photodetectors. This scheme of the measurements with the channels spatially separated along the front of wave allows one to use the film polarization elements like Polaroid films, decrease the optical path length, and use the noncoherent light source like a light-emitting diode with low power consumption.

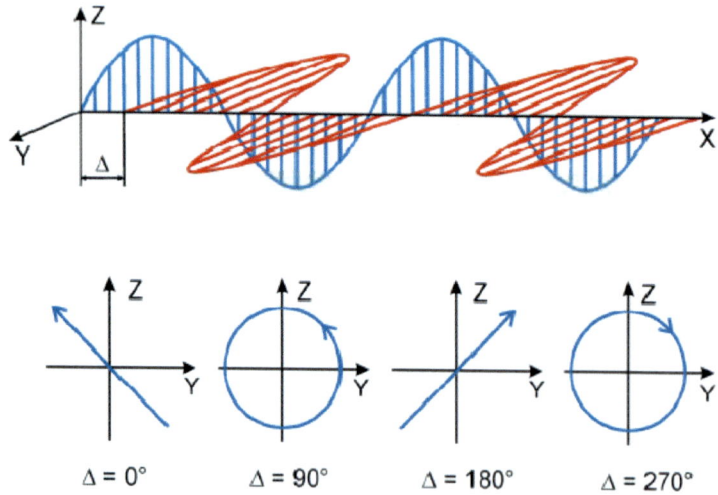

Figure 1.1. Space behavior of the electromagnetic wave (top) and particular cases of the resultant vector of the electric field intensity behavior (below).

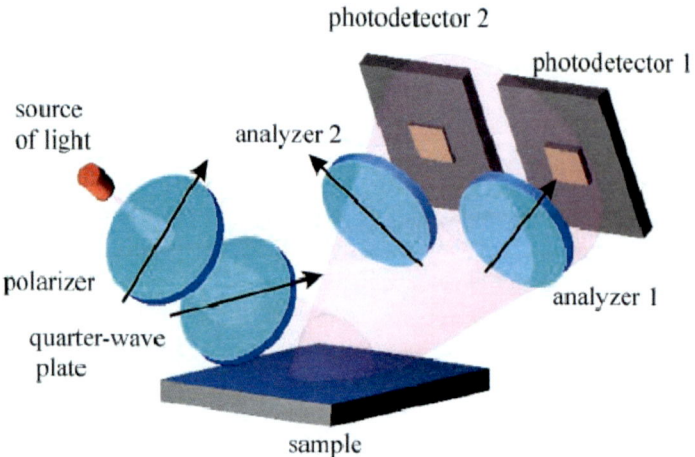

Figure 1.2. Optical scheme of the polarization-optical detector in the reflection mode. The arrows show the orientation of the optical elements axes.

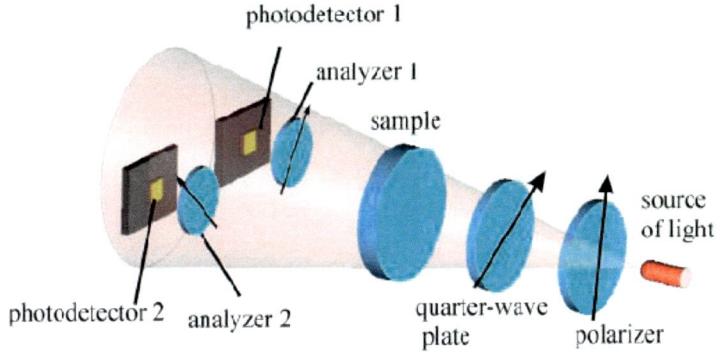

Figure 1.3. Optical scheme of the polarization-optical detector in the transmission mode. The arrows show the orientation of the optical elements axes.

The operation of the device consists in the following. It is well known that the magnitude of the intensity of light falling on the photodetector is determined by the orientation of the optical axes of the elements being part of the scheme [2]:

$$I = I_0\left[R_s^2 \cos^2(A) + R_p^2 \sin^2(A) + R_s R_p \sin(2A)\cdot\cos(2P - 90 + \Delta)\right]. \quad (1.1)$$

Here I_0 is the initial light intensity, P is the azimuth of the polarizer axis, A is the azimuth of the analyzer axis, R_s, R_p (or T_s, T_p) are the reflection coefficients (or transmission coefficients) mutually orthogonal components of polarization, and Δ is the phase shift between the components caused by the interaction with the sample. The Δ measurement is carried out at the initial orientation of the analyzers $A_1 = 45°$ and $A_2 = 135°$ and quarter-wave plate orientation $C = 45°$.

The equation (1.1) becomes:

$$I_1 = I_0\left[0.5\cdot(R_s^2 + R_p^2) + R_s R_p \cdot \cos(2P - 90 + \Delta_1)\right]$$
$$I_2 = I_0\left[0.5\cdot(R_s^2 + R_p^2) - R_s R_p \cdot \cos(2P - 90 + \Delta_2)\right] \quad (1.2)$$

The signal-processing system provides the measurement of differential signal I_Δ, where

$$I_\Delta = k \frac{I_1 - I_2}{I_1 + I_2} = k \frac{2R_s R_p}{R_s^2 + R_p^2} \cdot \sin\left(2P + \frac{\Delta_1 + \Delta_2}{2}\right) \cdot \cos\left(\frac{\Delta_1 - \Delta_2}{2}\right). \quad (1.3)$$

Here k is the proportional factor. Believing that parameters Δ_1 and Δ_2 are equal or the difference is lost in the noise, equation (1.3) becomes:

$$I_\Delta = k \frac{2R_s R_p}{R_s^2 + R_p^2} \cdot \sin(2P + \Delta). \quad (1.4)$$

It can be seen from equation that at $2P + \Delta = 90°$ the output signal changes δI_Δ will be proportional to the phase shift changes $\delta \Delta$:

$$\delta I_\Delta \approx \delta \Delta. \quad (1.5)$$

The sensitivity of the detector to the phase change makes 10^{-4} radian or 10^{-5} of wavelength. Thus, this scheme allowed one to develop the highly sensitive, small-size, light in weight, and low-power consumption devices for the measurement of various physical magnitudes described below.

Chapter 2

MEASUREMENT OF INFRA-RED RADIATION INTENSITY

OPTOACOUSTIC CELLS MATRIX STRUCTURE

The optoacoustic cells were used for the infra-red imager developing. The optoacoustic cell (OAC), or so-called Golay cell [3, 4], represents the hollow cylinder (Figure 2.1) filled with gas. One end face of the cylinder is an absorber of radiation and the opposite end face is the flexible membrane with a mirror-image covering. Sorption of radiation leads to the heating of the absorber and, then, the gas, which expands and deforms the membrane. The membrane deformation results in the deflection of the reflected beam of the visible light, which is detected by the photodetector.

In our work we have realized the *matrix* structure of the OAC for the imaging of IR radiation. The cell structure (Figure 2.2, on the left) is formed at an input window made of ZnSe, which is transparent in a range of wavelengths 0.5–20 microns. The diameter of the cell varied from 100 up to 200 microns and the height was 50 microns. The absorbing layer consisting of SiO_2 layer by thickness of 250 nanometers and aluminium layer by thickness of 10 nanometers settles down inside the cell. SiO_2 has an absorption band in the wavelengths range of 8–14 microns. The top view on the absorbing layer is shown in Figure 2.2 at the left below. The membrane consisted of polyimide and aluminium layers by the total thickness of 100 nanometers. The shaping of the matrix structure was carried out under the blanket of xenon with the purpose of filling the closed volume of optoacoustic cells with gas of low thermal conductivity. The fragment of the matrix structure is shown in Figure 2.3.

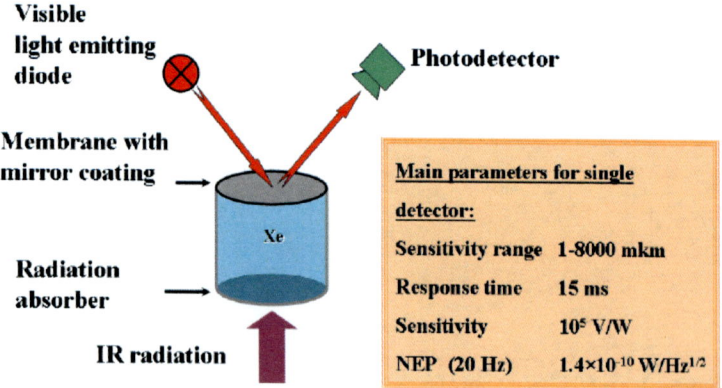

Figure 2.1. Functional diagram of the optoacoustic cell and the main parameters obtained for the single detector.

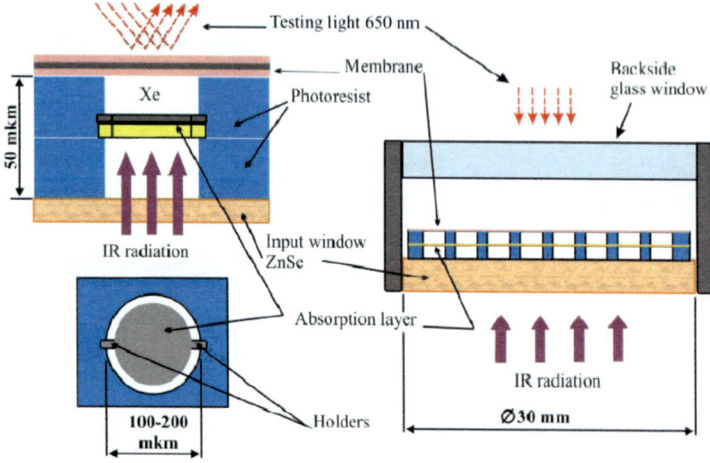

Figure 2.2. Scheme of the optoacoustic cell (left) and the matrix IR detector (right).

The resulted structure of optoacoustic cells matrix on input window ZnSe was located in the closed chamber of the IR matrix detector limited on one side by the ZnSe plate and, on the other side, by the glass plate (Figure 2.2, on the right). The given design automatically provided a regime of equality of an intrinsic pressure in the optoacoustic cell and an intrinsic pressure in the chamber at the change of ambient pressure and temperature.

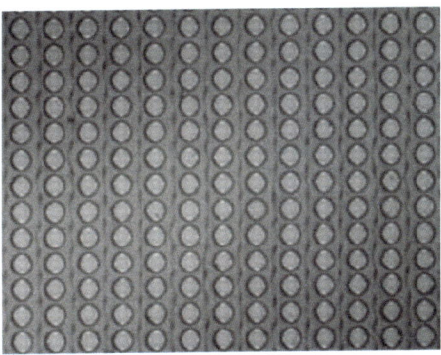

Figure 2.3. Fragment of the optoacoustic cells matrix structure.

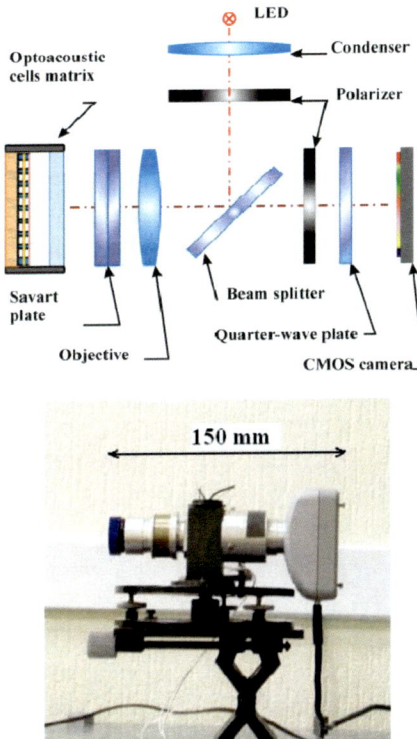

Figure 2.4. Scheme of the optical readout system (top) and the view of the IR imager (bottom).

OPTICAL READOUT SYSTEM

The scheme of the optical readout system and a view of the developed IR imager are shown in Figure 2.4 on the top and on the bottom, respectively. The light from a 650-nm-wavelength LED passes through the condenser, the polarizer, the beam splitter, the objective, and Savart plate, and it falls on the membrane surface of the optoacoustic cells. The membrane surface image is further projected onto the CMOS camera. The quarter-wave plate and the analyzer are established in front of the camera.

The key element of the system is Savart plate. The Savart plate consists of two identical plates (Figure 2.5) made of the uniaxial crystal (crystalline quartz). The plates are cut out in such a manner that the perpendicular to the plate surface makes the angle 45° with the optical axis of the crystal. From the position with identical orientation, the second plate is rotated around the perpendicular at the angle 90° related to the orientation of the first plate. At the arbitrary orientation of the input beam polarization, there are two linearly polarized rays spatially displaced after the first plate on size d due to double-refraction. Size d is defined by the difference of refraction indexes of quartz and by the plate thickness. After transition through the second plate, the rays are displaced in addition on size d in the direction perpendicular to the displacement after the first plate (as shown in Figure 2.5 below). As a result, the rays are displaced on size $\sqrt{2}\,d$ after the Savart plate.

The phase shift between the rays is zero because the plates are made on the uniform technology and have the identical thickness. Therefore, these rays are coherent. The given condition allows us to use noncoherent light sources. In our case the light-emitting diode with wavelength $\lambda = 650$ nanometers and emission line width $\Delta\lambda = 20$ nanometers was used.

The thickness of the Savart plate was chosen so that one of the rays hits the central part of the movable membrane and the second one hits the immovable place between membranes. In other words, $\sqrt{2}\,d$ is equaled to the membrane radius (Figure 2.6). The rays reflected from the membrane surface are focused on the CMOS camera surface and, as a result, some initial interference picture arises (Figure 2.6 on the left below). This picture is considered as the dark frame. The gas heating and expanding due to IR radiation sorption shifts the membrane from the initial point on Δx that leads to the phase shift between these rays and, as a result, to the change of the interference picture. Software of CMOS camera displays the change of the interference picture on a monitor related to the initial one. The differential interference image of the fragment of the optoacoustic cells matrix obtained by means of the described optical system is presented in Figure 2.6 on the right below.

The radiation, which has been passed through the Savart plate, reflected from the membrane surface, and passed again through the plate, can be presented in the form of

$$E_C = E_P \cdot \cos 45°$$
$$E_f = E_P \cdot \sin 45° \cdot \ell^{i\Delta}, \qquad (2.1)$$

where E_P is the amplitude of the polarized beam falling on the Savart plate after the polarizer (Figure 2.4) and the beam splitter; E_C and E_f are the amplitudes of the rays after the Savart plate; 45° is the angle between the direction of the initial polarization and the rays polarization directions after the Savart plate; Δ is the phase shift due to the membrane deformation.

The radiation falling on the quarter-wave plate (see Figure 2.4) can be decomposed on to two mutually perpendicular vectors conterminous in the direction with fast and slow optical axes of the quarter-wave plate. Considering that after the quarter-wave plate one of the component moves on the phase at 90°, we obtain the equations for amplitude of ray after quarter-wave plate:

$$E_{kno} = E_P \cdot (\cos 45° \cdot \cos \theta - \sin 45° \cdot \ell^{i\Delta} \cdot \sin \theta)$$
$$E_{kne} = E_P \cdot (\cos 45° \cdot \sin \theta + \sin 45° \cdot \ell^{i\Delta} \cdot \cos \theta) \cdot \ell^{i90}, \qquad (2.2)$$

where θ is the angle between ray polarization and quarter-wave plate axis.

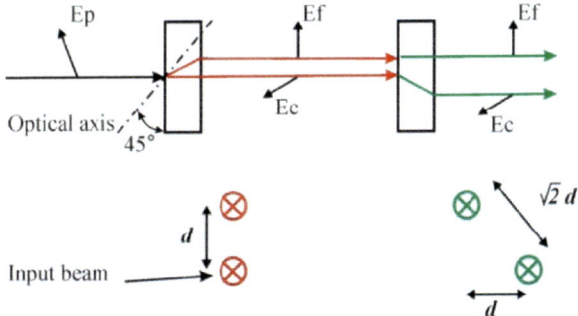

Figure 2.5. Scheme of the Savart plate.

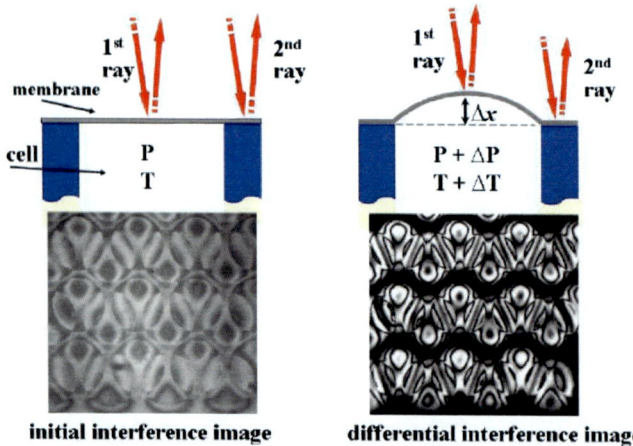

initial interference image differential interference image

Figure 2.6. Scheme of the phase shift between the reflected rays at the membrane deformation. The initial interference image (left below) and its change due to membrane deformation (right below).

The ray amplitude after analyzer (see polarizer in front of the camera in Figure 2.4) is defined by the equation:

$$E_A = E_{kno} \cdot \cos \varphi + E_{kne} \cdot \sin \varphi \qquad (2.3)$$

where φ is the angle of the analyzer orientation. The radiation intensity is defined as:

$$J = E_A \cdot E_A^*. \qquad (2.4)$$

Substituting values from equation (2.2) and believing $\theta = 45°$, we obtain:

$$J = [E_P \tfrac{1}{2} (1 - \ell^{i\Delta}) \cos \varphi + E_P \tfrac{1}{2} (1 + \ell^{i\Delta}) \ell^{i90} \sin \varphi] \times$$
$$\times [E_P \tfrac{1}{2} (1 - \ell^{-i\Delta}) \cos \varphi + E_P \tfrac{1}{2} (1 + \ell^{-i\Delta}) \ell^{-i90} \sin \varphi]. \qquad (2.5)$$

As a result, the radiation intensity at the certain point is

$$J = J_P \tfrac{1}{2} (1 + \cos(\Delta + 2\varphi)) \qquad (2.6)$$

It can be seen from equation (2.6) that the intensity depends on the phase shift, which is defined by the membrane deformation and the analyzer orientation. The maximal membrane deformation sensitivity is reached by installation of the analyzer in the position when $\Delta + 2\varphi = 90°$.

EXPERIMENTAL RESULTS

The sensitivity of the optical scheme to the membrane displacement, the membrane flexibility, and the own temperature sensitivity of the cell were defined. We accurately changed the pressure in the detector chamber and measured the signal amplitude at the certain point of the interference picture (square in Figure 2.7).

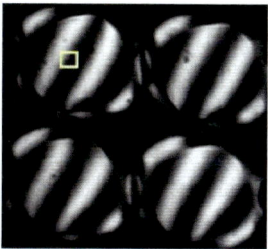

Figure 2.7. Differential image of the deformed membrane. The area where the signal was measured for the definition of the optical system sensitivity is shown by the square.

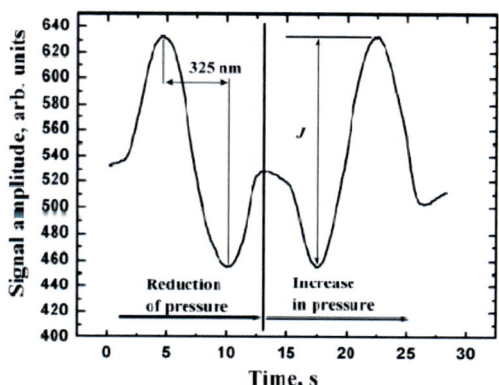

Figure 2.8. Time dependence of the signal amplitude from the optoacoustic cell at smooth reduction and then the increase in pressure in the detector chamber.

The time dependence of the signal amplitude from the optoacoustic cell at smooth reduction, and then the increase in pressure in the chamber of the IR matrix detector are shown in Figure 2.8. It can be seen from the figure and equation (2.6) that the change of the signal amplitude from the maximum up to the minimum (J) corresponds to the membrane displacement on $\lambda/2 = 325$ nanometers. The sensitivity of the optical scheme to the membrane displacement is defined as the amplitude of the membrane displacement Δx producing a signal-to-noise ratio of 1 and is determined as:

$$\Delta x = \frac{\Delta J}{J} \cdot \frac{\lambda}{2}, \qquad (2.7)$$

where ΔJ is the noise level. The displacement equivalent to noise takes the value of 1 nanometer.

The dependence of the membrane displacement from the pressure in the cell for various membrane diameters is shown in Figure 2.9.

The membrane flexibility is determined as the ratio $\Delta x/\Delta P$ where Δx is the membrane displacement and ΔP is the applied pressure. In other words, the membrane flexibility is defined by the slope of the dependences in Figure 2.9. It can be seen, that the flexibility increases by a factor of three at the twofold increase in diameter. Maximal sensitivity corresponding to the membrane with the diameter 205 nanometers takes the value of 0.5 nm/Pa.

The own temperature sensitivity of the cell can be evaluated from the minimal detected change in pressure ΔP. The time dependence of the optical signal at the periodic change of pressure in the detector gas volume on various magnitudes is shown in Figure 2.10. The minimal detected pressure was around $\Delta P = 2$ Pa.

It is well-known that the changes of temperature ΔT and pressure ΔP of gas at constant volume are connected with each other according to Mendeleyev-Clapeyron's equation:

$$\frac{P_0}{T_0} = \frac{P_0 + \Delta P}{T_0 + \Delta T}. \qquad (2.8)$$

Believing $P_0 = 100$ kPa and $T_0 = 300$ K, we obtain the own temperature sensitivity $\Delta T = 0.006$ K. Thus, the own sensitivity of the presented optoacoustic cell is close to a level of modern matrix thermal detectors [5].

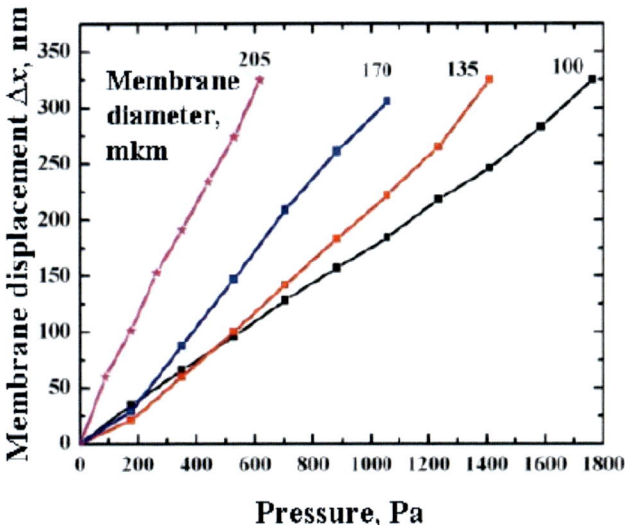

Figure 2.9. Membrane displacement dependence on the pressure in the cell for various membrane diameters.

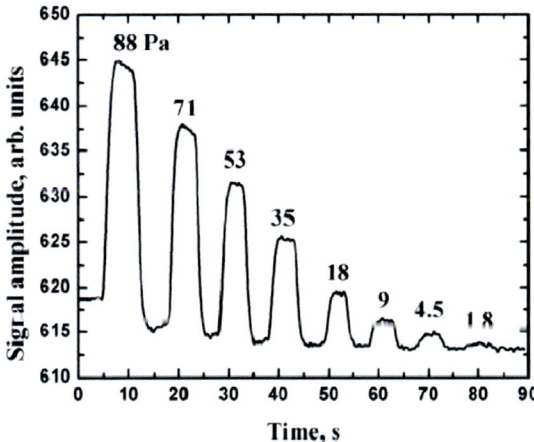

Figure 2.10. Time dependence of the optical signal at periodic change of pressure in the detector gas volume on various magnitudes.

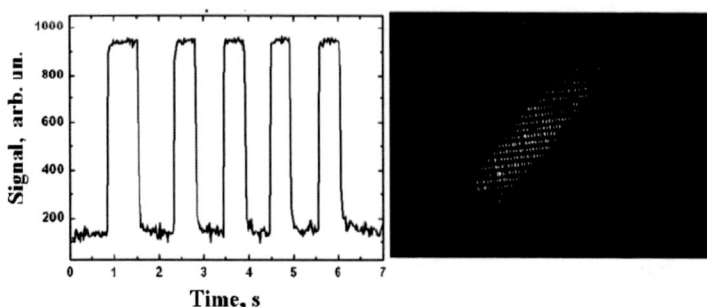

Figure 2.11. Time dependence of the signal response at the periodic exposure in the radiation of the blackbody at 200°C (left) and the image of an object heated at 200°C (right).

The signal response of the cell at the periodic exposure in the radiation from the blackbody at 200°C is shown in Figure 2.11 on the left. The response time was below 30 milliseconds. At 1 Hz and with optics f/1, the temperature resolution and noise equivalent power were 0.15 K and 11 nW, respectively. The image of an object at 200°C obtained by means of the 200×200 matrix optoacoustic cells detector is shown in Figure 2.11 on the right.

In the use of the optical readout method, there is no necessity to integrate a sensitive element and readout system in one crystal that is usual for semiconductor or bolometric matrix detectors [5]. It essentially simplifies manufacturing techniques and reduces the cost of the devices. Besides, such a detector is radiation-proof because it contains no radiation-unstable materials, and the image can be transferred by the fiber-optical communication channel. The separate systems of the detection and the reading are flexible to the sampling of the material for the input window and the absorptive layer that allows us to create both selective and broad-band detectors, which can operate from the ultraviolet through visible and infrared up to microwave wavelength as long as 8 mm.

Summing up, we can write down the transformations sequence from the signal amplitude to the measured physical magnitude:

$$\Delta I \propto \delta \Delta \propto \Delta x \propto \Delta P \propto \Delta T \propto \delta Q, \qquad (2.9)$$

where ΔI is the signal amplitude, $\delta \Delta$ is the phase shift change in the light wave, Δx is the displacement, ΔP is the change in pressure, ΔT is the change in temperature, and δQ is the radiation energy.

Chapter 3

MEASUREMENT OF THE ACOUSTIC SIGNAL SPECTRUM

THE PRINCIPLES OF THE ACOUSTIC SIGNAL ANALYSIS

Systems of the spectral analysis of acoustic signals find increasing application in the acoustic diagnostics of mechanical devices, speech recognition systems and sonar devices. Existing analyzers of the acoustic signal spectral composition commonly use the specialized digital methods like Fast Fourier Transform (FFT) [6–8], to which the signal recording, during certain time (Δt), precedes (Figure 3.1). The spectral resolution of the Fourier transform strongly depends on the ratio time duration/wavelength of the analyzed signal. The good spectral resolution can be reached only at high enough ratios. Therefore, the given method is mainly applied to the analysis of the established oscillations weakly varying in time. The given registration method has a low accuracy at the pulse acoustic signal when the act of the signal does not have regular character in time and makes some periods (e.g., human speech, sonar pulse).

In the present contribution, the multiple-unit structure of resonant sensors based on the production engineering of micro-electromechanical systems (MEMS) [9-11] with the optical readout system is offered to use for the analysis of the spectral composition of acoustic vibrations. The structure represents a set of microstrings having various resonant oscillation frequencies. The oscillation of each of strings arises when the analyzed signal contains the frequency conterminous with the resonant frequency of the string. Thus, each string is the original filter, which is adjusted on certain frequency. The reading of string oscillation amplitude is carried out by the optical method with the use of the commercial CMOS camera. In this system the signal-reading is carried out by the well developed and low cost multiplex system of the camera. The given registration system allows one to define the acoustic signal spectral composition in the parallel regime (Figure 3.1) that provides high rapidity and accuracy of the analysis. The design of the string structure allows us to create the systems with sensitivity up to ultrasonic range that is important for hydrolocation because the sound is a unique way of long distance reconnaissance and communication under water, sound emitter location and recognition [12–14].

The multi-strings resonant sensor is similar in the principle of its action with the human ear cochlea, where there is a spatial division of the acoustic signal along the basilar membrane, and registration of a signal is carried out by the series of sensory hair cells along the basilar membrane responding to send the neural pulses towards the brain [15]. Note that the human (or mammal) ear is the best acoustic vibrations analyzer for today.

THE STRUCTURE OF MECHANICAL RESONANT SENSORS

The tungsten strings 15–80 microns in diameter were used as the acoustic resonating elements in the analyzer structure. The strings were coiled on a rigid tungsten frame with controlled tension and fixed then by pressing with the subsequent high-heat treatment. The schemes and overviews of the fabricated string analyzers of the round, elliptical and trapezium forms are shown in Figure 3.2.

Measurement of self-resonant string frequencies of the acoustic sensor was spent by the excitation of the oscillations by the electric field with the use of the piezoelectric oscillator. The quantity of string resonators on one sensor structure made up 60–200, the frequency range of 100–20000 Hz, the half-width of the resonance line was 5 Hz that is close to human ear resolution.

Measurement of the Acoustic Signal Spectrum 19

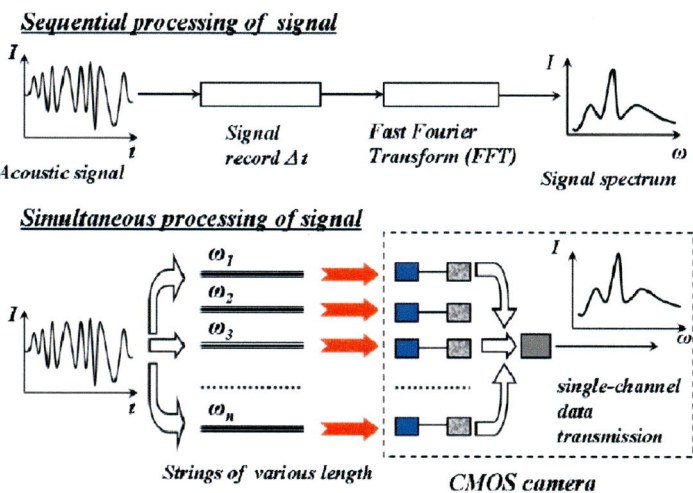

Figure 3.1. Schemes of the sequential and the simultaneous signal processing.

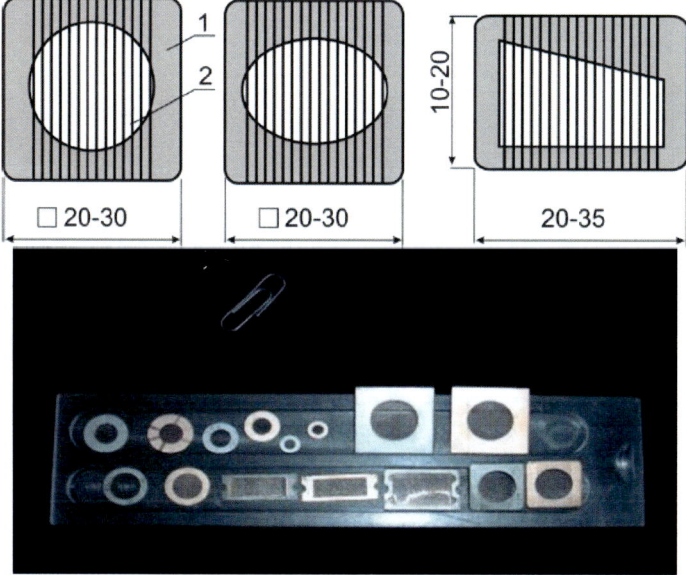

Figure 3.2. Schemes of the acoustic detector structures (top): 1, tungsten frame; 2, the strings. Overview of the fabricated string structures (below).

METHOD OF THE ACOUSTIC VIBRATIONS MEASUREMENT

The excitation of string resonant oscillations occurs on frequencies multiple to the reference frequency. For example, the resonance of a string observed at frequency ω_s = 40 Hz can arise under the act of frequencies $\omega_k = k \cdot \omega_s$, k =1, 2, 3... The given condition does not allow interpreting unambiguously the results of measurements. For the unique definition of the signal frequency, the procedure of mixing the measured signal with a signal of the fixed frequency (modulating frequency) ω_0 was carried out. As a result, the resonance arises only at frequencies ω_s, ω_0, and $\omega_0 \pm \omega_s$. The frequency ω_0 was chosen so that the summed up frequency $\omega_0 + \omega_s$ fell only into the range of analyzer sensitivity (Figure 3.3). That allows us to uniquely define the frequency of the measured acoustic signal.

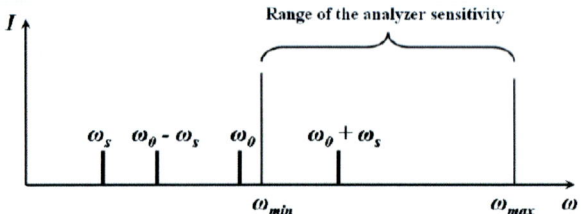

Figure 3.3. Scheme of the modulating frequency selection. ω_0, modulating frequency; ω_s, signal frequency.

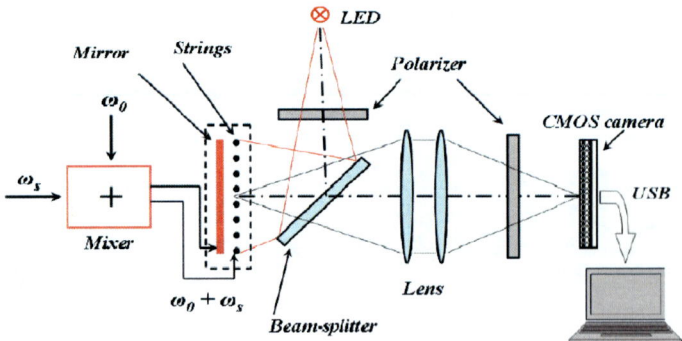

Figure 3.4. Scheme of the optical readout system.

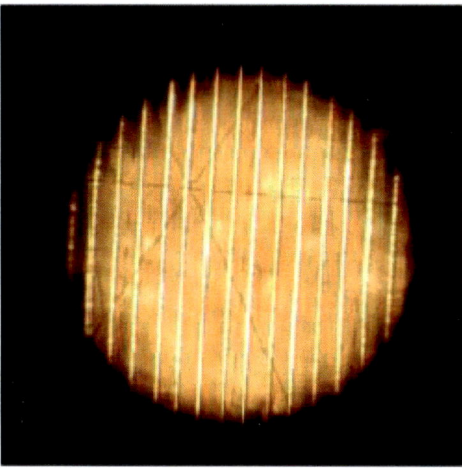

Figure 3.5. Image of the strings obtained with CMOS camera.

The scheme of the optical readout system is shown in Figure 3.4. The visible light from the light-emitting diode passes through the polarizer and the beam splitter falls on the string structure of the acoustic sensor. The light reflected from strings and metallic reflection surface passes through the beam-splitter, objective, and the polarizer and falls on the CMOS camera surface. The objective projects the strings image onto the CMOS camera surface. The optical axes of polarizers are turned to each other by 90°. That provides the quenching of specular reflected components and the alleviation of a spurious light act. The fragment of the string structure image is shown in Figure 3.5.

The main problem of the definition of the magnitude of the string oscillation amplitude is that the matrix element acquisition interval is not less than 40 ms (25 Hz), whereas the string oscillations period lies over the range 10–0.05 ms (0.1–20 kHz). Hence, it is impossible to define directly the oscillation amplitude by means of the direct transformation of the optical signal to the electric signal using the CMOS matrix element. Therefore, the mathematical model, which described the magnitude of the electric signal of the matrix element at the variable optical signal with the period much smaller than the acquisition interval was developed.

The intensity distribution of the light reflected from the unexcited string and projected in the plane of CMOS matrix can be described by the function of Gaussian type (Figure 3.6):

$$E_1 = E_0 \exp\{-(mx)^2\}, \tag{3.1}$$

where x is coordinate perpendicular to the string line and m is a constant (Figure 3.6). The light intensity created by the vibrating string is described by the equation:

$$E_2(x,t) = E_0 \exp\{-(m(x + \Delta x \sin \omega t))^2\}, \tag{3.2}$$

where Δx is amplitude of string oscillation (Figure 3.6), ω is oscillation frequency, and $x = 0$ is the equilibrium point of the string under the absence of oscillations. The magnitude of the signal from the matrix element is the result of integration of equation (3.2) on the period equal to the acquisition interval (40 ms) at $x = 0$:

$$I_2 = E_0 \int \exp\{-(m\Delta x \sin \omega t)^2\} dt . \tag{3.3}$$

Solution of the given equation is possible by function $\exp\{-z^2\}$ decomposition in a row:

$$\exp\{-z^2\} = 1 - \frac{z^2}{1!} + \frac{z^4}{2!} + ... \tag{3.4}$$

and by consideration only the first two terms. Equation (3.3) becomes:

$$I_2 = E_0 \int_{t_1}^{t_2} (1 - m^2 \Delta x^2 \sin^2 \omega t) dt , \tag{3.5}$$

where $\Delta t = t_2 - t_1$ is the matrix element acquisition interval. The result of the integration gives the magnitude of the signal from the matrix element measured in the point $x = 0$:

$$I_2 = E_0 (1 - \frac{1}{2} m^2 \Delta x^2) \Delta t . \tag{3.6}$$

The signal difference between the unexcited and excited string is

$$\Delta I = I_1 - I_2 \approx \Delta x^2. \qquad (3.7)$$

Δx^2 is proportional to vibrational energy, which is, in turn, proportional to the energy of the analyzed acoustic signal at the frequency equal to that string resonant. Thus, it becomes possible to unambiguously connect the magnitude of the optical signal with the string oscillation amplitude.

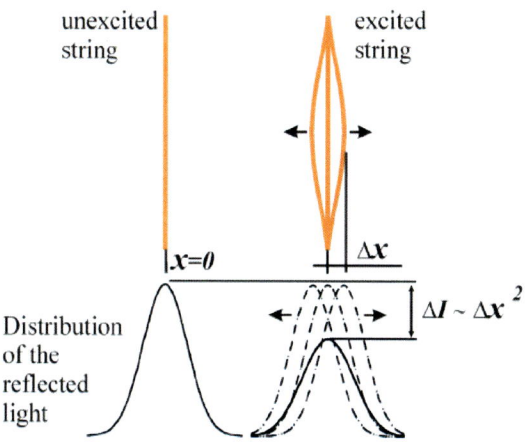

Figure 3.6. Model of the distribution of the intensity of the light reflected from unexcited (left) and excited (right) strings. The solid lines show the light intensity averaged through the matrix element acquisition interval.

Summing up, we can write down the transformations sequence from the signal amplitude to the measured physical magnitude:

$$\Delta I \propto \Delta x^2 \propto Q(\omega), \qquad (3.8)$$

where ΔI is the signal amplitude, Δx is the amplitude of string oscillation, and $Q(\omega)$ is the acoustic signal intensity at frequency ω.

EXPERIMENTAL RESULTS

The diagram of the intensity of the light reflected from the string structure is shown in Figure 3.7 (top). The number of the element in the matrix row is shown on x-coordinate, and signal intensity in arbitrary units is shown on *y*-coordinate. Narrow peaks correspond to strings. Wide peaks correspond to nonuniformity of the light intensity on the structure area. The diagram of the difference of two frames obtained at the exposition 40 ms is shown in Figure 3.7 below and reflects the noise level.

The diagram of the difference of the frames between excited and unexcited states of the resonant sensor is shown in Figure 3.8. The signals with frequency ω_s =119 Hz (measured signal) and frequency ω_0 = 4200 Hz (modulating frequency) are given to the mixer inputs. It can be seen from the figure that the resonances arose only at frequencies corresponding to ω_0 and $\omega_0 + \omega_s$. The diagram of the signal spectrum with the casual set of frequencies is shown in Figure 3.9.

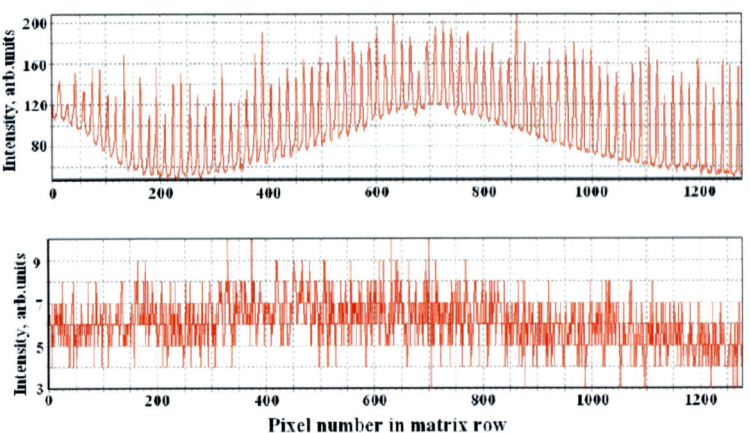

Figure 3.7. Diagram of the light intensity from the string structure (top) and the diagram of the difference of two sequentially obtained frames showing the noise level (below).

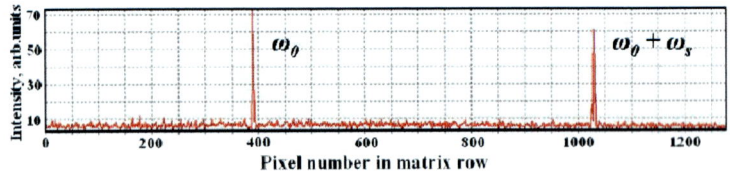

Figure 3.8. Diagram of the signal spectrum with the signal frequency ω_s =119 Hz and the modulating frequency ω_0 =4200 Hz.

Measurement of the Acoustic Signal Spectrum

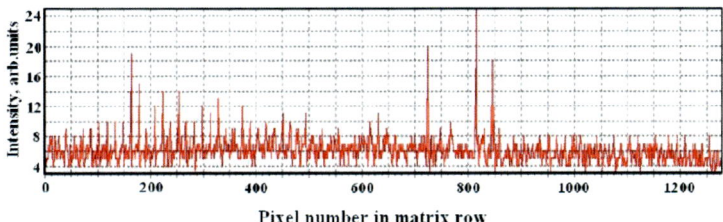

Figure 3.9. Diagram of the random signal spectrum.

The given techniques allow us to create small-size rapid-transfer spectrum analyzers of acoustic oscillations for speech recognition systems and sonar devices.

Chapter 4

MEASUREMENT OF THE DEFORMATION

The piezo-optical effect was used for the development of high-sensitive linear strain (deformation) gauges.

THEORETICAL BACKGROUND

The piezo-optical effect (alternatively called the photoelastic effect) is the change of refractive index caused by stress [16, 17]. The piezo-optical effect is observed in all crystal and amorphous substances. The effect is noticeably shown in glass, sintered quartz, silicon, and germanium. The use of photoelasticity involve applying a given stress state to a model and utilizing the induced birefringence of the material to examine the stress distribution within the model. Photoelasticity remains a major tool in modern stress analysis.

The applied stress results in a change in the refractive index of a transparent substance due to deformation of the atomic electron shells. If a general system of stresses is applied in a plane, Δn produced by optical birefringence becomes proportional to the difference $\Delta\sigma$ between two principal stresses in the plane [18].

$$\Delta n = n_o - n_e = K \Delta\sigma. \tag{4.1}$$

Here n_o and n_e are the refractive indexes for ordinary and extraordinary rays, respectively, and K is the stress-optical coefficient with typical orders of magnitudes $10^{-11} \div 10^{-12}$ m^2/Newton for most optical materials. The difference

Δ between phases of ordinary and extraordinary rays having passed through an optical material by thickness l is:

$$\Delta = \frac{2\pi l}{\lambda}(n_o - n_e) = c\Delta\sigma l, \qquad (4.2)$$

where $c = 2\pi K/\lambda$ is a new constant. The constant c can be positive or negative depending on the material.

Taking into account that the sensitivity of the developed detector to the phase shift is 10^{-4} radian and using the Hooke's law, we can estimate the sensitivity of the strain gauge to the deformation.

$$\Delta\sigma = E\frac{\Delta x}{x} \Rightarrow \Delta = \frac{2\pi l}{\lambda}KE\frac{\Delta x}{x} \Rightarrow \frac{\Delta x}{x} = \Delta\frac{\lambda}{2\pi l K E}. \qquad (4.3)$$

Here E is the modulus of elasticity; x is the size of the photoelastic element; Δx is the deformation. For the sintered quartz, this is used as the photoelastic element, $E = 70$ GPa, $K = 1.4 \times 10^{-12}$ m^2/Newton [19]. Believing that $l = 4$ mm, $\lambda = 650$ nm, $x = 10$ mm, the sensitivity to the deformation takes the value of $\Delta x \approx 2.7 \times 10^{-10}$ m = 0.27 nm.

Summing up, we can write down the transformations sequence from the signal amplitude to the measured physical magnitude:

$$\Delta I \propto \delta\Delta \propto \Delta n \propto \Delta\sigma \propto \Delta x, \qquad (4.4)$$

where ΔI is signal amplitude, $\delta\Delta$ is phase shift change in the light wave, Δn is change in refractive index, $\Delta\sigma$ is stress, Δx is displacement.

STRAIN GAUGE AND TENSOMETRIC STATION

The scheme and overview of the developed strain gauges and its location modes are shown in Figure 4.1. The gauges of this design are intended for the stress measurement in all directions in the sample plane.

The sintered quartz was used as a photoelastic specimen; the polarizers and analyzers were made of Polaroid films. The base size of the gauge is 15 mm, the range of the linear deformation measurement is 10 microns with sensitivity 0.001 micron. The linear dynamic range of the output signal is

10^3. The achieved sensitivity is 50 times as higher as the sensitivity of the film resistor gauges. The unique property of sintered quartz consists in the absence of the residual effects after multiple applied stresses, which exceed the dynamic range by a factor of 30.

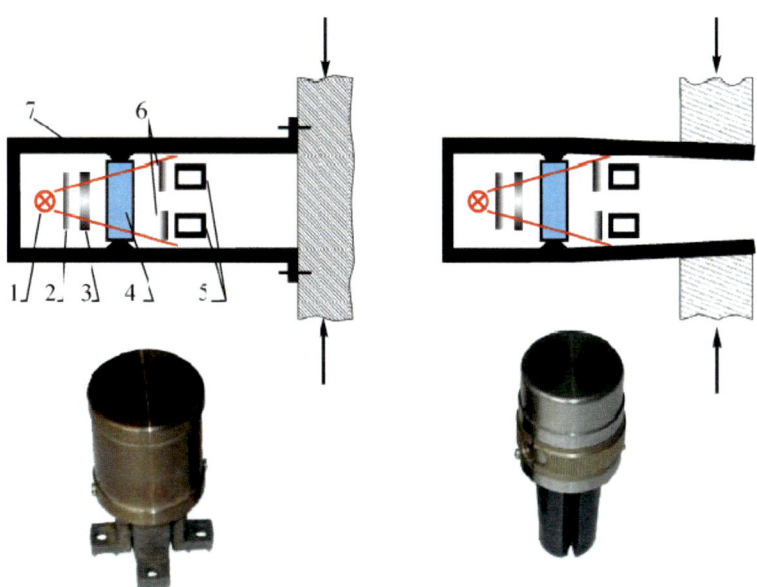

Figure 4.1. Scheme and location modes of the optical strain gauge. *1*, LED; *2*, polarizer; *3*, quarter-wave plate; *4*, photoelastic element; *5*, photodetectors; *6*, analyzers; *7*, metal frame. The arrows show the direction of the stress.

Actually the range of deformation measurements can be adjusted to the desired magnitude by changing the mechanical design of the gauge. The accuracy of the measurements is 0.01% of the dynamic range. The linear strain gauge was also developed (Figure 4.2). The base size of the linear stress gauge is 12 mm, the range of the linear deformation measurements is 10 mm with the accuracy 0.001 mm. In this design the stress can be measured in only one direction.

The linear strain gauges were used for the development of the strain probe designed for the detection of the lateral rock deformation in relief holes with small-scale diameter (80–85 or 115–120 mm) in three directions by means of linear strain gauges shifted by 120 degrees (Figure 4.3).

Figure 4.2. Overview of the optical linear strain gauge.

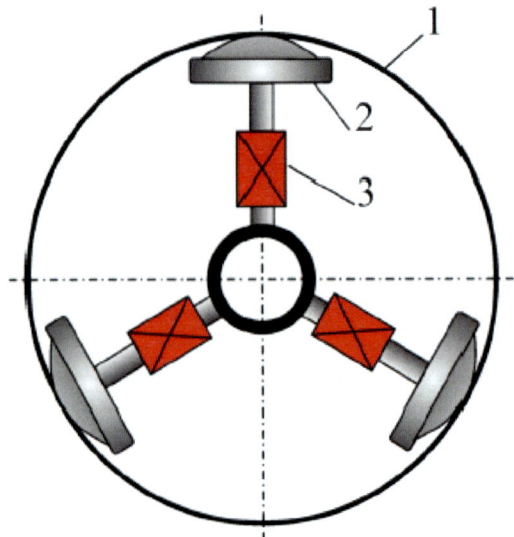

Figure 4.3. Scheme of the strain probe. *1*, wall of the hole; *2*, tips which set against the wall, and *3*, linear strain gauges.

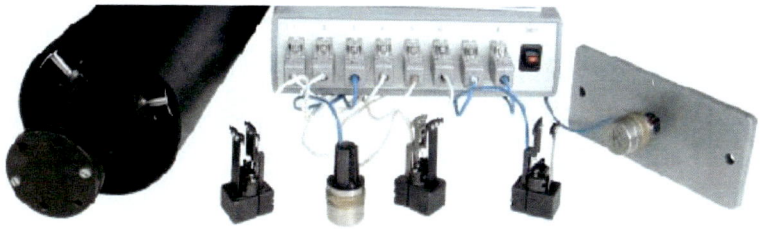

Figure 4.4. Tensometric station.

The tensometric station with the set from 1 to 8 of the optical strain gauges is designed for the investigation of the strain distribution in the details and construction (Figure 4.4). That allows online monitoring the shearing of load-carrying frame at rock bursts, earthquakes, and dangerous technical activities and can be used in engineering industry, mineral resource industry, metallurgy, building industry, and secure facilities.

Chapter 5

MEASUREMENT OF THE PHYSICAL AND CHEMICAL PARAMETERS OF LIQUID AND GAS

The effect of the total internal reflection was used for the development of gauges for the measurement of physical and chemical parameters of liquid, gas, and biological environments. The developed refractometers for the gas and liquid analysis are described in this section, and the DNA sensor — in the next section.

THEORETICAL BACKGROUND

Total internal reflection is an optical phenomenon that occurs when a ray of light strikes a substance interface at an angle larger than the critical angle with respect to the normal to the surface. If the refractive index is lower on the back side of the interface, no light can pass through, and the light is totally reflected. The critical angle is the angle of incidence, above which the total internal reflection occurs.

When light crosses an interface between materials with different refractive indexes, the light beam is partially refracted at the interface surface, and is partially reflected (Figure 5.1). However, if the angle of incidence is greater (i.e. the ray is closer to being parallel to the interface) than the critical angle, i.e., the angle of incidence at which light is refracted so that it travels along the interface, then the light stops crossing the interface and, instead of that, is totally reflected back internally. This can occur only where light travels from a substance with a higher refractive index to one with a lower refractive index. For example, it occurs when light passes from glass to air but does not when from air to glass. The critical angle φ_c is given by the Snell's law:

$$\varphi_c = \arcsin\left(\frac{n_2}{n_1}\right), \tag{5.1}$$

where n_2 is the refractive index of the less optically dense substance and n_1 is the refractive index of the optically denser substance.

There are two aspects of total internal reflection. The first is that the reflected light has an angle dependent phase shift between the reflected and incident light. This phase shift depends on polarization and increases as the incidence angle deviates further from the critical angle toward grazing incidence. The second aspect is the propagation of an evanescent wave across the boundary surface. Actually, even though the entire incident wave is reflected back into the originating substance, there is some penetration into the second substance at the interface. Both these aspects allow one to design the optical devices for the investigation of very thin layers (~λ) close to the interface.

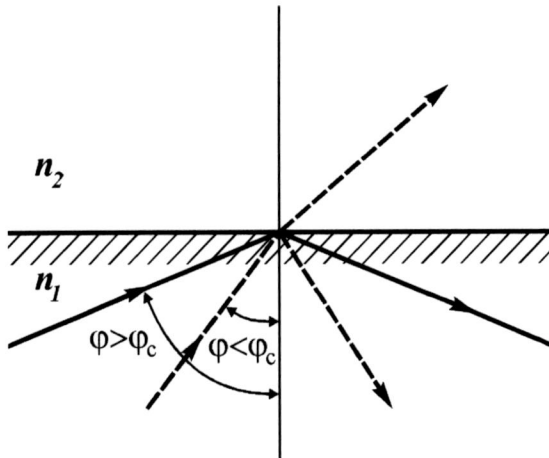

Figure 5.1. Scheme of the total internal reflection. The larger the angle to the normal, the smaller the fraction of light is transmitted; total internal reflection occurs when the angle achieves the value of φ_c.

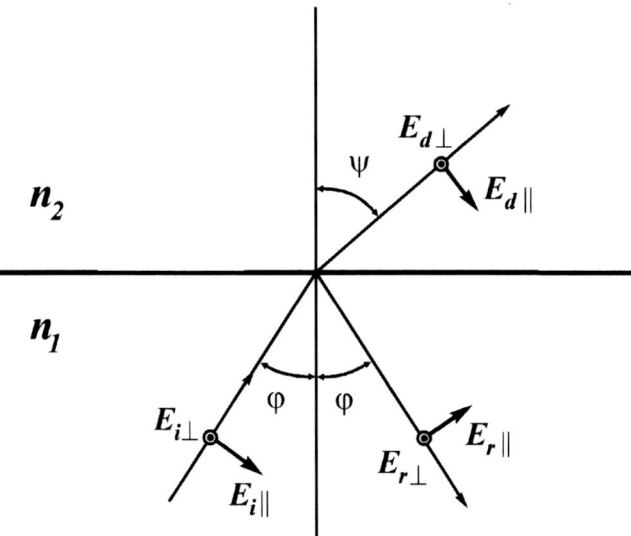

Figure 5.2. Scheme of the electric field vectors position in the incident E_i, reflected E_r, and refracted E_d electromagnetic waves. The electric field vector components $E_{j\|}$ lie in the incidence plane and components $E_{j\perp}$ are perpendicular to the incidence plane.

The amplitudes of the reflected and refracted waves are determined by the Fresnel equations [20]. If the incident light is polarized with the electric field of the light in the plane of the diagram (Figure 5.2), the amplitude ratios for the reflection and the transmission are:

$$r_\| = \frac{E_{r\|}}{E_{i\|}} = -\frac{\tan(\varphi - \psi)}{\tan(\varphi + \psi)}, \qquad (5.2)$$

$$t_\| = \frac{E_{d\|}}{E_{i\|}} = \frac{2\sin\psi \cos\varphi}{\sin(\varphi + \psi)\cos(\varphi - \psi)}. \qquad (5.3)$$

If the incident light is polarized with the electric field of the light perpendicular to the plane of the diagram (Figure 5.2), the amplitude ratios for reflection and transmission are:

$$r_\perp = \frac{E_{r\perp}}{E_{i\perp}} = -\frac{\sin(\varphi-\psi)}{\sin(\varphi+\psi)}, \qquad (5.4)$$

$$t_\perp = \frac{E_{d\perp}}{E_{i\perp}} = \frac{2\sin\psi\cos\varphi}{\sin(\varphi+\psi)}. \qquad (5.5)$$

Here E_i, E_r, E_d are amplitudes of the incident, reflected, and refracted waves, respectively; φ and ψ are angles of the incidence and refraction, respectively. According to the Snell's law, the ratio between refractive indexes is:

$$\frac{\sin\varphi}{\sin\psi} = \frac{n_2}{n_1} \equiv n_{12}. \qquad (5.6)$$

It can be seen from equation (5.6) that at $n_{12} < 1$ φ can take values, at which $\sin\psi > 1$ that is not meaningful. This takes place for all values of φ, which meet the conditions $\sin\varphi > n_{12}$. That is possible only for $n_{12} < 1$. Since the angle ψ is not meaningful at these conditions, we cannot use the Fresnel equations because ψ is their immediate constituent. But we can transform these equations by introducing n_{12}. Let us replace $\sin\psi$ by $\sin\varphi / n_{12}$ and $\cos\psi$ by $\pm\sqrt{1-\sin^2\varphi/n_{12}^2}$. For the case considered here, $\sin\varphi / n_{12} > 1$. That means that $\sin^2\varphi / n_{12}^2 > 1$, i.e. $\cos\psi$ becomes the imaginary argument:

$$\cos\psi = \pm i\sqrt{\frac{\sin^2\varphi}{n_{12}^2} - 1}. \qquad (5.7)$$

The plus sign corresponds to infinite amplitude increase as moving away the reflecting interface to the second substance. Therefore, the final equation becomes:

$$\cos\psi = -i\sqrt{\frac{\sin^2\varphi}{n_{12}^2} - 1}. \qquad (5.8)$$

After calculations we have E_r and E_d expressed in terms of E_i, φ, and n_{12}. However, calculated expressions are not real but complex. Complex amplitude values for reflected and refracted waves have a simple meaning:

the argument of the complex amplitude defines the phase shift between the incident and the reflected (refracted) waves. The components $E_{r\perp}$ and $E_{r\parallel}$ change the phase relatively to the $E_{i\perp}$ and $E_{i\parallel}$, which are denoted as $\Delta_\perp$ and $\Delta_\parallel$, respectively, and can be expressed as:

$$\tan\frac{\Delta_\parallel}{2} = -\frac{\sqrt{\sin^2\varphi - n_{12}^2}}{n_{12}^2 \cos\varphi}, \quad \tan\frac{\Delta_\perp}{2} = -\frac{\sqrt{\sin^2\varphi - n_{12}^2}}{\cos\varphi} \quad (5.9)$$

The $\Delta_\perp$ differs from $\Delta_\parallel$, and we can write down the equation for the relative phase difference $\Delta = \Delta_\perp - \Delta_\parallel$ as:

$$\tan\frac{\Delta}{2} = \frac{\tan\frac{\Delta_\perp}{2} - \tan\frac{\Delta_\parallel}{2}}{1 + \tan\frac{\Delta_\perp}{2}\tan\frac{\Delta_\parallel}{2}} = \frac{\cos\varphi\sqrt{\sin^2\varphi - n_{12}^2}}{\sin^2\varphi}. \quad (5.10)$$

Hence, if $E_{i\perp}$ and $E_{i\parallel}$ in the incident wave are in one phase, the phase shift Δ appears between $E_{r\perp}$ and $E_{r\parallel}$ in the reflected wave, which depends from φ and n_{12}. Therefore the primordially linear polarized wave is transformed to the elliptically polarized wave after the total internal reflection. The phase shift changes $\delta\Delta$ caused by the changes in refractive index δn_{12} (due to chemical reaction, adsorption-desorption processes, etc. in the substance with n_2 (see Figure 5.2)) at fixed φ are defined by the equation:

$$\frac{\delta\Delta}{2\cos^2\frac{\Delta}{2}} = -\frac{\cos\varphi}{\sin^2\varphi}\frac{2n_{12}\delta n_{12}}{\sqrt{\sin^2\varphi - n_{12}^2}}, \quad (5.11)$$

where $n_{12} = n_2/n_1$, and $\delta n_{12} = \delta n_2 / n_1$. It can be seen from the equation that $\delta\Delta \sim \delta n_{12}$.

This is the base of the developed sensors for the refractive index changes measurements.

At the total internal reflection the evanescent wave penetrates only 100-300 nanometers into the media with lower refractive index and its intensity (I) at a given depth (z) can be calculated as:

$$I(z) = I(0)\ell^{-z/d}, \qquad (5.12)$$

where:

$$d = \frac{\lambda}{4\pi}\left[n_1^2 \sin^2\varphi - n_2^2\right]^{-1/2} \qquad (5.13)$$

and λ is the wavelength of the light impinging on the interface at angle φ. The effective penetration depth is equal to $\sim \lambda$. This yields a very small investigated volume near the interface and allows us to develop various research techniques like the total internal reflection fluorescence microscopy [21]. The significance of this illumination configuration is that it greatly reduces the background signal typically associated with the inhomogeneity of the investigated layer in thickness.

Summing up, we can write down the transformations sequence from the signal amplitude to the measured physical magnitude:

$$\Delta I \propto \delta\Delta \propto \Delta n, \qquad (5.14)$$

where ΔI is signal amplitude, $\delta\Delta$ is phase shift change in the light wave, and Δn is the refractive index change caused by the change in physical and chemical properties of substance.

THE LIQUID REFRACTOMETER

In this contribution the electron refractometers were developed for the refractive index measurements of liquid solutions. The flowing (Figure 5.3) and contact (Figure 5.4) detectors were designed. The 650-nm LED is used as a light source, the accuracy of the refractive index measurements being 5×10^{-5}. The range of the refractive index measurements depends on the refractive index of the prism and varies from 1.33 to 1.49.

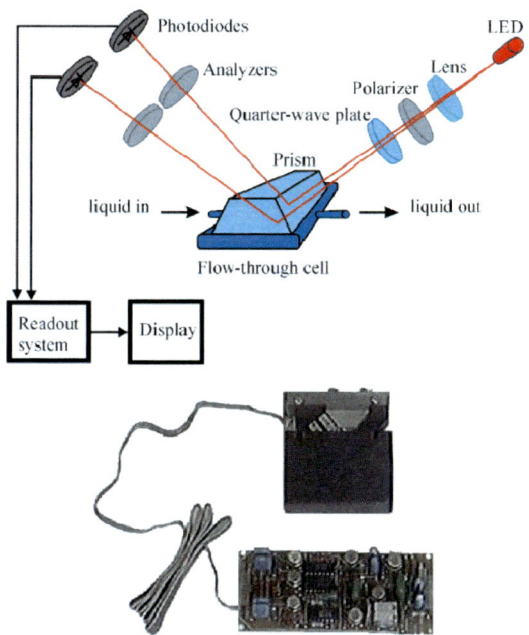

Figure 5.3. Scheme and overview of the flowing liquid refractometer.

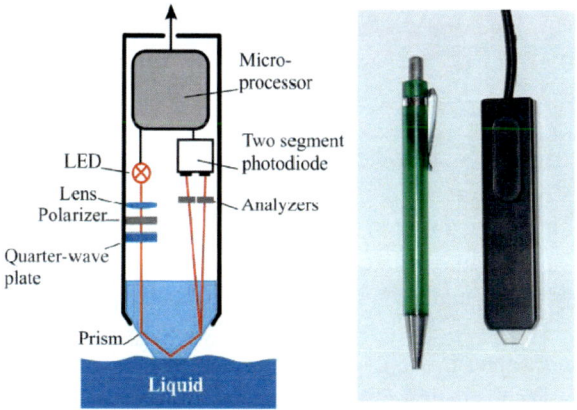

Figure 5.4. Scheme and overview of the contact liquid refractometer.

These refractometers can be used as gauges in liquid chromatography and in automatic test equipment for the production quality control in food and chemical industries.

THE GAS SENSOR

The optical scheme of the gas sensor is quite the same as for liquid refractometer (Figure 5.5) except the gas volume. The method of the gas analysis is based on the measurements of the changes in the refractive index of the absorbed film during the interaction with molecules of gas. The absorbed film has a contact with prism, in which the total internal reflection conditions occurs. The chemical reaction between gas molecules and receptor molecules of the absorbed film leads to the change in molecular polarizability and, as a result, to the change in refractive index. There are two modes for the refractive index change measurements depending on the investigation task: 1) the total internal reflection at the absorbed film–gas interface; 2) the total internal reflection at the prism–absorbed film interface. The choice of the mode is defined by the prism refractive index and the incidence angle of the light ray on the prism. The first mode is more suitable for the surface adsorption-desorption processes; the second one, for the bulk diffusion processes.

The refractive index of a substance consisting of R types of molecules is conditioned by molecular polarizability according to the Lorentz–Lorenz equation [20]:

$$\frac{n^2-1}{n^2+2} = \frac{4\pi}{3}\sum_R N_R \alpha_R, \qquad (5.15)$$

where n is the refractive index, N_R is the number of molecules with the polarizability α_R per unit volume, and R is the number of types of molecules (bonds).

Let us assume that M new molecules with polarizability α_Q are generated in absorbed film due to the chemical reaction of molecules of R type with gas molecules. Refractive index n_M of the film with a new composition is defined from equation:

$$\frac{n_M^2 - 1}{n_M^2 + 2} = \frac{4\pi}{3}\left[\sum_{R-1} N_{R-1}\alpha_{R-1} + (N_R - M)\alpha_R + M\alpha_Q\right] \quad (5.16)$$

The refractive index changes ($\Delta n = n - n_M$) caused by this chemical reaction can be determined by equation:

$$\frac{6\Delta n \cdot n}{(n^2 + 2)^2} = \frac{4\pi}{3} M(\alpha_R - \alpha_Q). \quad (5.17)$$

Note that if α_R and α_Q are known, the number M can be evaluated.

The developed gas sensor has the sensitivity to the phase shift (Δ) ~ 2×10^{-3} degrees, linear dynamic range $\Delta \pm 10$ degrees, sensitivity to the refractive index change Δn ~ 5×10^{-5}, size $42\times25\times13$ mm^3, weight 50 grams. Output signal is proportional to the phase shift:

$$\Delta I = K \cdot \Delta, \quad (5.18)$$

where $K = 500$ mV/degree.

Summing up, we can write down the transformations sequence from signal amplitude to the measured physical magnitude:

$$\Delta I \propto \delta\Delta \propto \Delta n \propto M, \quad (5.19)$$

where ΔI is signal amplitude, $\delta\Delta$ is phase shift change in the light wave, Δn is refractive index change, and M is the number of new molecules generated during the chemical reaction.

The results of sulfur dioxide detection are presented as an example in Figure 5.6 and Figure 5.7 [22]. The polysiloxane films modified by alkylamines were used as the absorbed film. This material was chosen as the sensitive film because sulfur dioxide forms compounds of donor-acceptor type with alkylamines.

The total internal reflection occurred at the interface between absorbed film and gas as most sensitive mode to the surface absorption. The sharp increase of the signal is observed during the first few seconds. Then the signal rise becomes slower but the full saturation is not observed even in 50 minutes of exposition (Figure 5.6, inset). At the loss of SO_2 supply, the signal sharply goes down and returns to the initial value during less than 1 minute. Therefore, the interaction process of SO_2 with the sensitive film is fully reversible. It can be seen from Figure 5.6 that desorption rate exceeds significantly the adsorption rate. The signal saturation, which is characterized by the establishment of the adsorption-desorption equilibrium, is not achieved. That means that the diffusion of the SO_2 molecules into the bulk of the sensitive film takes place along with adsorption onto the surface.

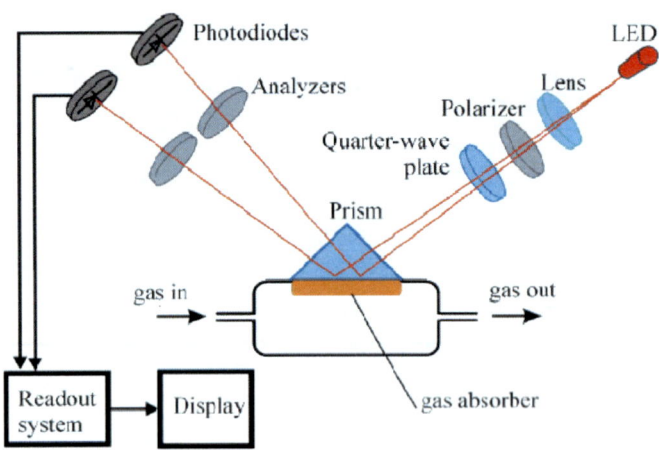

Figure 5.5. Scheme of the optical gas sensor with adsorbed film.

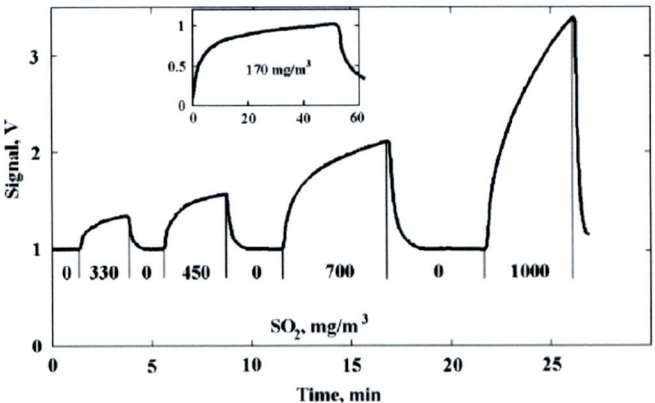

Figure 5.6. Time dependence of the signal (proportional to Δn) at the periodic exposure to SO_2 at various concentrations. The long time exposition at SO_2 concentration 170 mg/m^3 is shown in the inset.

It is impossible to determine the SO_2 concentration by the signal amplitude at the absence of the signal saturation. Therefore, the signal change rate (V/min) at the beginning of the SO_2 exposure was chosen as the static characteristic of the sensor. The dependence of the signal change rate from SO_2 concentration is quite linear up to concentration of 1000 mg/m^3 (Figure 5.7). The deviation from linearity does not exceed 5 %. The detection limit takes the value of 20–30 mg/m^3.

The described principle for measurements of SO_2 concentration can be used for the development of various gas sensors. The proposed sensor can be used for both the investigation of the surface and bulk diffusion processes in the sensitive film and kinetics of chemical reactions. This sensor has high sensitivity, small size, and low energy consumption, commensurable with semiconductor sensor devices and allows the implementation of the non-contact method of the measurements.

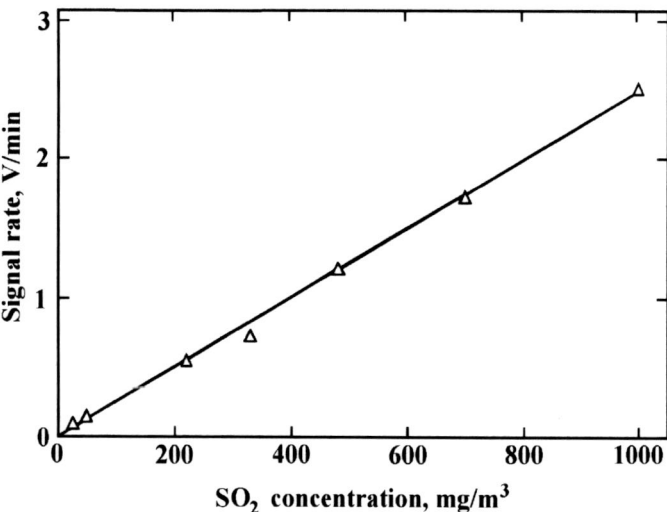

Figure 5.7. Dependence of the signal change rate on SO_2 concentration in the beginning of SO_2 exposure (see Figure 5.6).

Chapter 6

THE DNA SENSOR

PROBLEMS AND PRINCIPLES OF THE DNA ANALYSIS

The diagnostics of social important diseases such as hepatitis, AIDS, tuberculosis, etc., is a fundamental problem of great value in all over the world. The revealing of hereditary, infectious, and oncological diseases requires reliable, accurate, and specific methods for the DNA analysis. DNA is a double helical in nature and consists of two strands, which contain four nucleotides as monomer units (A, T, G, and C) (Figure 6.1). The double helix is formed due to the so-called hybridization process, i.e., complementary sequence-specific interactions between base pairs from neighboring strands: *Adenine-Thymine, Guanine-Cytosine.* Short DNA fragments, oligonucleotides consisting of up to 40 nucleotide units are widely used for both the investigation in molecular biology and designing test systems for DNA diagnostics due to their ability to form complementary complexes with nucleic acids. The stage of hybridization of nucleic acids with oligonucleotide probes, i.e. the formation of complementary complexes, ensures the specificity and selectivity of the analysis.

The modern technologies based on the method of molecular hybridization widely use the so-called DNA chips or DNA arrays which are solid supports containing a set of immobilized oligonucleotide probes of defined sequences [23, 24]. DNA chips are used for the detection of the interaction between an oligonucleotide probes on the chip surface with complementary regions of an analyzed DNA. If DNA from the analyzed sample contains the region complementary to the immobilized oligonucleotide probe, it forms a stable complex while contacting with the chip surface. DNA molecules containing no such sequences are washed from the surface (Figure 6.2).

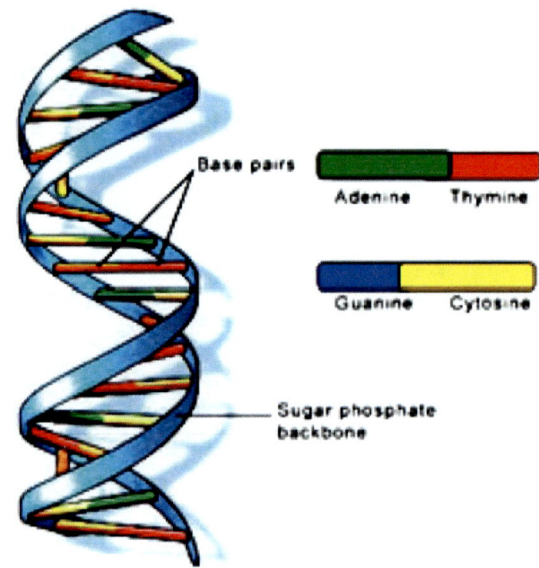

Photo courtesy U.S. National Library of Medicine.

Figure 6.1. Fragment of DNA double helix.

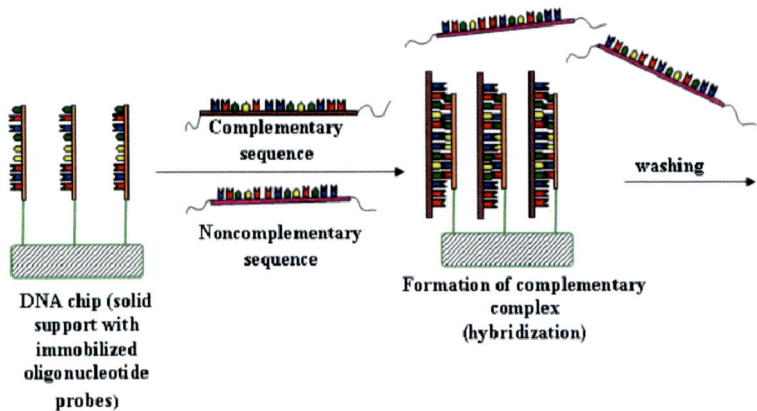

Figure 6.2. Scheme of hybridization between an analyzed DNA sample and an oligonucleotide probe immobilized onto a solid support.

DNA from an analyzed sample is amplified by means of the polymerase chain reaction (PCR) and, at the same time, fluorescent or another label is introduced into the PCR fragment. The resulted DNA molecules are loaded onto the chip. The attachment of this DNA to complementary oligonucleotide probes leads to the appearance of the label in definite zones. Registration of the labeled zones allows the conclusion that specific DNA in the analyzed sample is present.

Complementary complexes can be registered by the so-called DNA sensors using the physical methods for the detection of the selective sorption of certain molecules on solid supports [25–27]. The optical and electrical characteristics of films immobilized onto the surface can vary depending on the processes on the film/medium interface, in particular, on the sorption-desorption processes. Various physical methods of the detection can be exploited in DNA sensors: quartz crystal microbalance (the oscillation frequency of the crystal containing immobilized oligonucleotides changes after the sorption of complementary DNA strands) [28, 29], surface plasmon resonance (the refractive index of the metal surface changes while the ligand from the solution binds to molecules immobilized onto the metal) [30, 31], scanning force microscopy [32, 33], X-ray photoelectron spectroscopy [34, 35]. In electrochemical detection, electrodes with immobilized oligonucleotide probes change their electrical characteristics during the hybridization process [36, 37]. Optical sensors detect changing the optical properties of immobilized films (e.g., refractive index or interference) in hybridization processes. The above methods provide the high sensitivity and allow the analyses to be carried out more rapidly, simpler, and more efficiently as compared to convenient methods. Their use in medical practice, however, is restricted because the sensors are too costly to produce for the commercial market.

DNA CHIP AND SENSOR

In this work a polarization refractometry sensor, which is intended for the hybridization DNA analysis is described. The operation principle of the sensor is based on measuring the refractometric characteristics of the sensitive surface layer (see section 5). The proposed sensor can be used for the detection of complementary complexes formed between oligonucleotide probes immobilized on the surface of a DNA chip and analyzed DNA fragments. The interaction of molecules from the solution with the receptors on the surface leads to the formation of new complexes and, thereby, to the change of optical characteristics of immobilized films caused by changing their structure and thickness (see equations 5.15–17 in section 5). Registration of the resulted process of the DNA sorption is performed by measuring the difference of phases (Δ) between two orthohonal components of light wave arising at the total internal reflection on the film-medium interface (see equations 5.10, 5.11 in section 5).

Scheme and overview of the sensor are shown in Figures 6.3. and 6.4.

The proposed DNA sensor is completed with a DNA chip that is a prism of length 2 cm and the base angle 53° made of glass with the refractory index 1.74. The prism surface is covered by the film consisting of three layers SiO_2-ZrO_2-SiO_2 (280-120-190 nm, respectively) or one TiO_2 layer (300 nm) by means of the electron-beam evaporation or magnetron sputtering methods, respectively. The resulted surface is used for immobilizing oligonucleotide probes of various sequences corresponding to different analyzed DNAs. The scheme of the chip is shown in Figure 6.5.

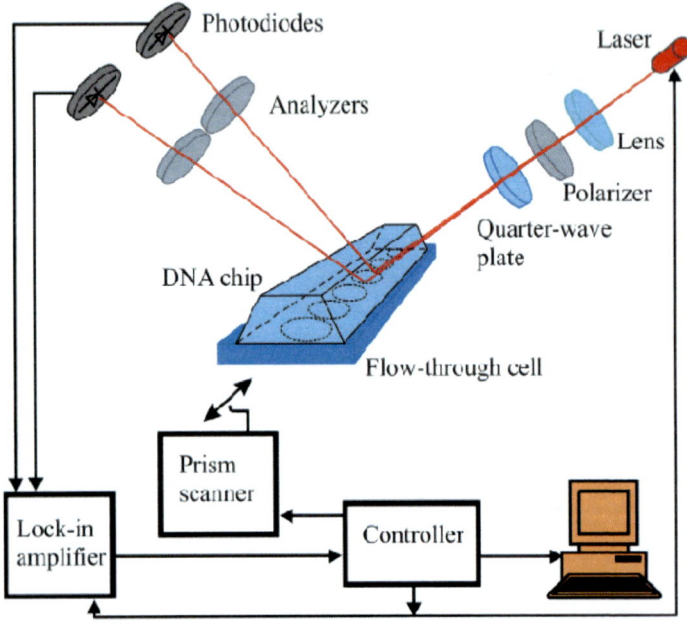

Figure 6.3. Optical and operational scheme of DNA sensor.

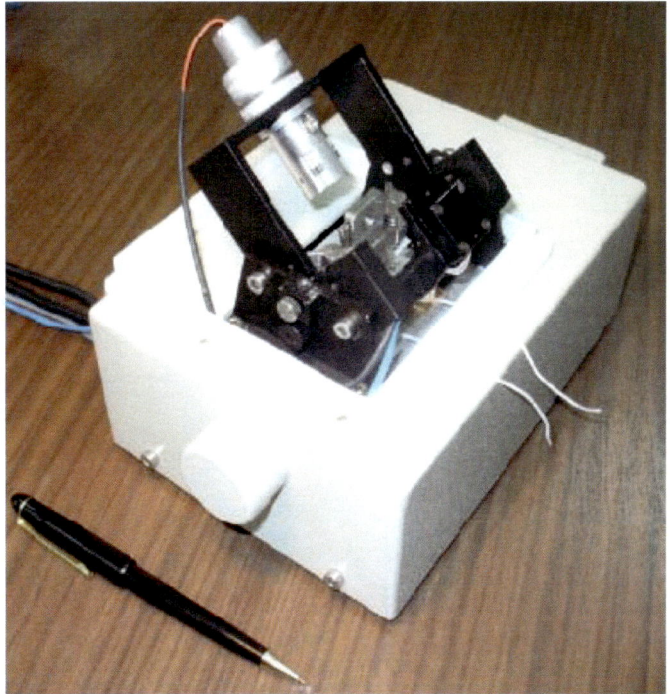

Figure 6.4. Overview of DNA sensor in the DNA chip scanning regime.

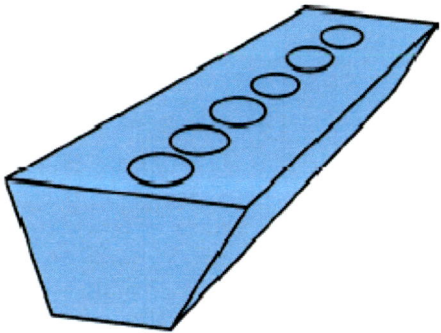

Figure 6.5. DNA chip: glass prism containing immobilized oligonucleotide probes.

Oligonucleotides were immobilized by the method based on the use of polylysine-containing oligonucleotide derivatives [38]. The loading of oligonucleotides (10 to 15 probes, ≈ 0.5 mm in diameter) on the prism surface were carried out in the sensor body supplied by the special holder and scanning arrangement ensuring the high accuracy of the prism positioning (<1 microns) so that oligonucleotides were placed in defined positions along the straight line (Figure 6.6).

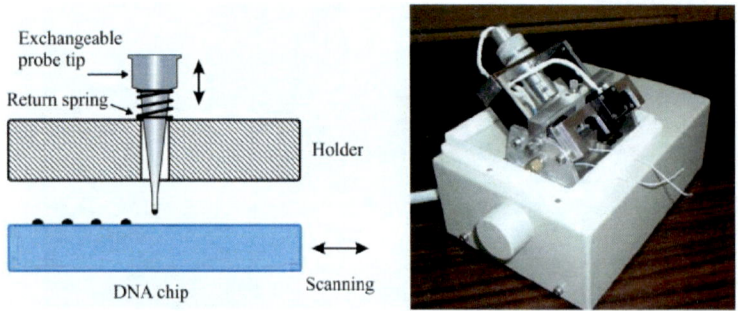

Figure 6.6. Scheme of the loading of the probes on the chip surface and overview of the DNA sensor in the probe loading regime.

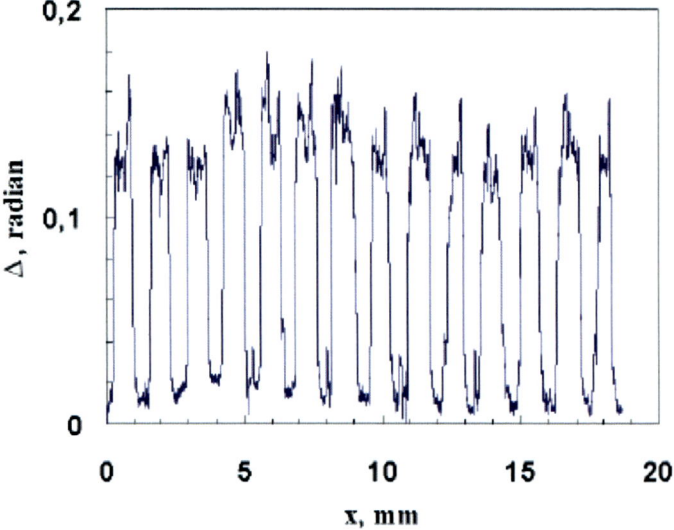

Figure 6.7. Scan of the prism containing immobilized oligonucleotide probes. X is distance along the chip; Δ is the phase shift in the reflected light along the chip. Peaks correspond to the signal from probes.

The scanning of the prism in measuring regime was carried out using the same scanning arrangement that allowed one to significantly enhance the signal/noise ratio and, thereby, the reliability of the measuring results. After the probes were spotted, the prism was washed from non-bound oligonucleotides and prepared for measurements. The resulted DNA chip, after drying, was placed in the special holder so that the volume of the reaction chamber (flow-through cell) was 4 microliters. The immobilization result can be monitored by scanning the prism in "dry condition" before the following reactions (Figure 6.7).

The prism containing immobilized probes was washed from the unbound oligonucleotides and subjected to the hybridization reaction with an analyzed DNA sample, which were labeled with the biotin residue ensuring the higher sensitivity of the analysis. After hybridization, the prism was treated with streptavidin-alkaline phosphatase conjugate (SA-AP) (Figure 6.8) in order to increase the amount of substance precipitated on the chip surface and to increase, thus, the signal intensity (see equation 5.17 – 5.19 in section 5).

If the oligonucleotide probe immobilized on the prism was complementary to the region of an analyzed biotinylated DNA sample, they were hybridized, and the biotin label was bound to the surface. Streptavidine from the SA-AP conjugate forms an extremely tight complex with the biotin label and the conjugate was also attached to the surface [39]. All the above reactions were fulfilled outside the DNA sensor. The prism prepared by such a way was then placed inside the sensor, and the subsequent reaction of alkaline phosphatase with chromogenic substrates (Figure 6.7) was carried out under the sensor's observation. Chromogenic substrates and washing buffers were injected into the flow-rate reaction cell with a syringe. The reaction of alkaline phosphatase with chromogenic substrates leads to the formation of a sediment strongly at the place of the biotin label allocation. The sensor registers signals at these places where hybridization with an analyzed complementary biotinylated DNA fragments takes place. The biotin label was also inserted into the control oligonucleotide probe used as a positive control (+P), so the signal was also registered where this control probe is located. The scanning of the prism in the measuring regime was carried out using the same scanning arrangement as in the loading regime that allowed one to significantly enhance the signal/noise ratio and, thereby, the reliability of the measuring results.

The total internal reflection on the prism/solution interface was used to avoid the influence of buffer and reagent solutions on the measurement (see section 5). In order to eliminate the influence of the surface inhomogeneity, the differential method was used. The initial state of the surface is detected before the reaction with chromogenic substrates (zero reading), then the surface state is detected after the reaction. The differential signal defines the areas where the complementary oligonucleotide complexes are generated or the positive control is placed.

After measurements, the prism can be cleaned with a soft sponge wetted with alcohol, washed with water, cleaned with ultra sound and be ready to a new cycle. Thus, the proposed DNA chips are applicable for multiple tests.

Summing up, we can write down the transformations sequence from signal amplitude to the result of the analysis:

$$\Delta I \propto \delta\Delta \propto \Delta n \rightarrow +P(-P), \tag{6.1}$$

where ΔI is the signal amplitude, $\delta\Delta$ is the phase shift change in the light wave, Δn is the refractive index change, $+P$ $(-P)$ is positive (negative) reaction to the presence (absence) of the sought DNA.

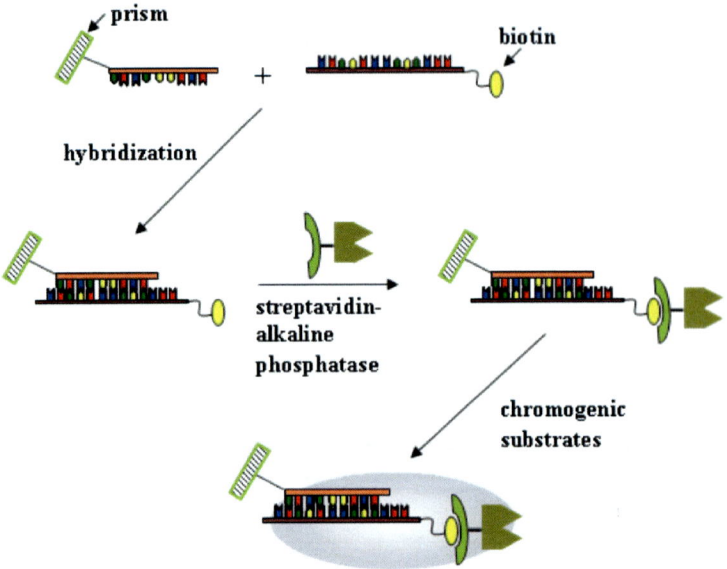

Figure 6.8. Scheme of successive reactions on DNA chip: hybridization of an analyzed biotinylated DNA sample with an oligonucleotide probe, complex formation between the biotin label and streptavidine from the streptavidin-alkaline phosphatase conjugate, and reaction of alkaline phosphatase with chromogenic substrates.

EXPERIMENTAL RESULTS

Prisms with three-layer film (SiO_2-ZrO_2-SiO_2) were tested in experiments on hybridization of the model biotinylated analyzed sample (AS-bio) corresponding to nucleotide sequences of E gene from tick-born encephalitis virus (TBEV) with the complementary oligonucleotide probe immobilized on the prism surface. The prism containing immobilized complementary and noncomplemenatry probes and positive control was submitted to the reaction with AS-bio followed by the reaction with the SA-AP conjugate. The prism was then inserted into the sensor, scanned in a buffer solution, treated with chromogenic substrates, and scanned again in 30 min after the beginning of the reaction. The result of the difference between these two scans is presented in Figure 6.9. Signals are registered at 1–4 and 9–10 positions where complementary probe (c-P) and positive control (+P) are located, respectively. Signals at 5–8 positions with noncomplementary probes (n/c-P) are hardly detected.

After the addition of chromogenic substrates to the biotinylated complementary complex formed while hybridizing of oligonucleotide AS-bio with the immobilized complementary probe, the prism was scanned at certain time intervals (Figure 6.10, on the left). The kinetic curve of the process is shown in Figure 6.10 on the right. The intensity of signals, when revealing spots with chromogenic substrates, achieves the plateau values in 30–40 min after the beginning of the reaction.

One of the main requirements for immobilized films of oligonucleotide probes on the DNA chip surface is its homogeneity. It is especially important for DNA sensors with the optical detection methods because it affects the noise level and the accuracy of the polarization state analysis. In addition to prisms with a three-layer film consisting of silicon and zirconium dioxides, in this contribution we also tested the titanium dioxide film with the thickness of 300 nm vacuum-deposited by magnetron-sputtering source. Oligonucleotide spots on such surface after drying of the spotting solution form a more homogeneous film than in case of silicon dioxide (Figure 6.11).

Prisms with the TiO_2 film containing the immobilized oligonucleotide probes were tested in experiments on hybridization with the model oligonucleotide fragment AS-bio. The prism contained the spots of immobilized complementary and noncomplemenatry probes and positive control. After hybridization with AS-bio and reaction with the SA-AP conjugate, the prism was placed into the sensor, treated with chromogenic substrates, and scanned in 30 min after the beginning of the reaction. It is seen from Figure 6.12 that the signals are registered only in the case of complementary probes (positions 4, 5, 7, 8, 10, 11) and positive control (positions 1, 2, 13, 14). Signals at places of noncomplementary probe coincide with the noise level.

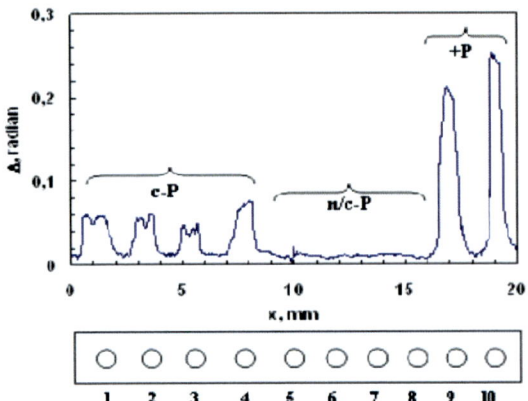

Figure 6.9. Result of hybridization of biotinylated oligonucleotide (AS-bio) (10^{-10} molar concentration) corresponding to nucleotide sequences of E gene from TBEV with complementary oligonucleotide probe immobilized on the prism. Scanning the prism was carried out before and after the revealing of spots with chromogenic substrates. The resulted picture is the difference between these two registrations. Complementary (c-P) and non complementary (n/c-P) probes and positive control (+P) are immobilized on the prism surface at positions 1–4, 5–8, and 9–10, respectively.

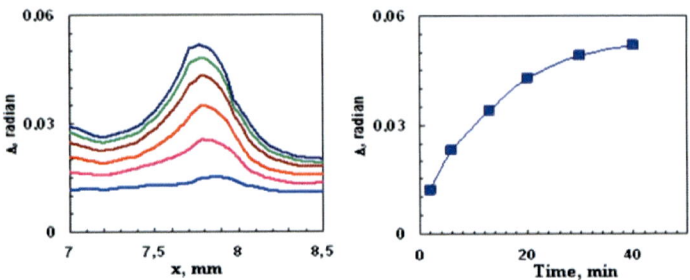

Figure 6.10. Fragment of the prism scan during the revealing of spots with chromogenic substrates in 2, 6, 13, 20, 30, and 40 min after the beginning of the reaction (left); kinetic curve of the reaction registered at x = 7.8 mm (right). The revealing of spots was carried out after hybridization of biotinylated oligonucleotide AS-bio (10^{-10} molar concentration) with immobilized complementary oligonucleotide probe (c-P).

The DNA Sensor 55

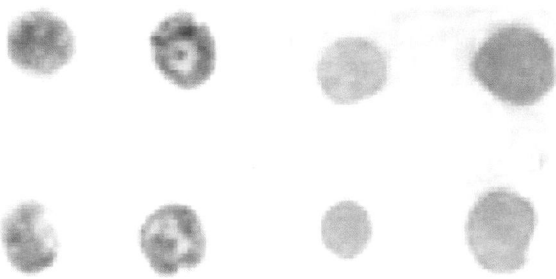

Figure 6.11. Spots of oligonucleotide bearing the biotin label revealed with the SA-AP conjugate and chromogenic substrates on the SiO_2 surface (left) and on the TiO_2 film (right).

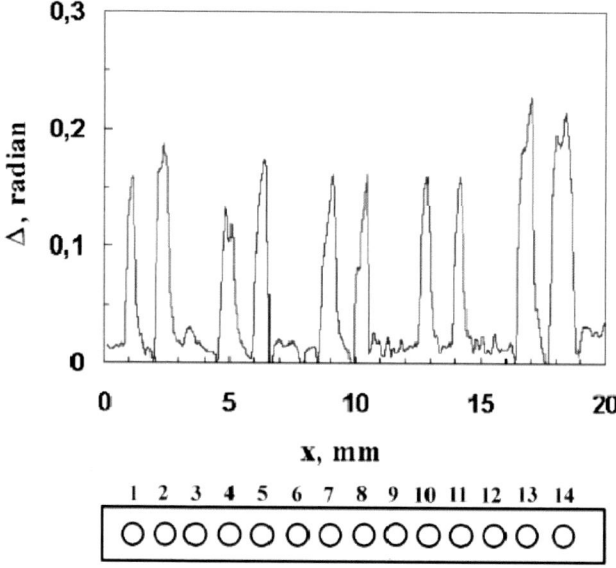

Figure 6.12. Result of hybridization of biotinylated oligonucleotide AS-bio (10^{-10} molar concentration) with complementary oligonucleotide probe on the prism surface. Scanning the prism was carried out before and after the revealing of spots with chromogenic substrates. The resulted picture is the difference between these two registrations. Complementary probes (c-P) were immobilized at 4, 5, 7, 8, 10, and 11 positions; non complementary probes (n/c-P), at 3, 6, 9, and 12 positions; positive control (+P), at 1, 2, 13, and 14 positions.

It should be noted that prisms covered with TiO_2 films provide the higher sensitivity of the signals as compared to the previously used three-layer films (SiO_2-ZrO_2-SiO_2) (compare Figures 6.9 and 6.12). Thus, TiO_2 covering is preferable for at least, two reasons: it ensures the higher sensitivity of signals and higher homogeneity of immobilized oligonucleotide spots.

Polarization refractometry DNA sensor proposed in this contribution is intended to register the formation of complementary DNA complexes and, thereby, to reveal certain sequences in the DNA structure. Thus, the sensor can be applied for the rapid mass analysis of nucleic acids, namely, for diagnostics of various diseases; it can also be useful for optimization of the treatment regime. The use of DNA sensors makes it possible to obtain quantitative characteristics of processes upon hybridization of nucleic acids. The development of instrumental diagnostic methods designed to reduce the influence of subjective factors when identifying and treating various diseases. In a broader sense, the DNA sensors can be employed in the detection of biological warfare agents and in forensic science for the investigation of crime scenes, etc.

Chapter 7

CONCLUSION

Interaction of light with substance is caused by the interaction of electromagnetic field of the light wave with electromagnetic field of the molecules forming a given substance (solid, liquid, gas). This interaction leads to the change in the state of polarization of the light wave. This provides the unique possibility to investigate the substance in noncontact regime. In this work, the optical scheme and the method for the measurement of the state of the light wave polarization with a split-hair accuracy are developed. This method utilizes two measuring channels spatially separated along the front of the wave. It allows using the film polarization elements like Polaroid films, decreasing the optical path length and using noncoherent light source like light-emitting diode. On the basis of this optical scheme, the small size and low power consumption liquid refractometers, gas, strain and DNA sensors were designed. They are intended for measurements of the liquid refractive index, kinetics of the chemical reactions in liquids and gasses, stresses and deformations, gas detection. The DNA sensor is designed for the analysis of viral or bacterial nucleic acids responsible for dangerous diseases.

The optical scheme, where the spatial separation of light beam on two coherent rays is carried out by the Savart plate, is used for the development of the matrix IR radiation imager. The optical readout system detects the membrane displacement with an accuracy of 1 nm resulting from radiation absorption by the receiving element. That provides the own temperature sensitivity below 0.006 K. Thus, using the developed optical readout system, the inexpensive and compact wide-range thermal imager can be created as well as other devices, where the high-displacement sensitivity is required.

The optical readout system was used for the development of the acoustic signal analyzer. The analyzer consists of the set of microstrings of various length and diameter and having resonant frequencies over the range of 100–20000 Hz. The half-width of the string resonance line was 5 Hz, which is close to human ear resolution. The developed method for the string oscillation amplitude measurement with the use of the commercial CMOS camera allows us to define the acoustic signal frequency spectrum with high rapidity and accuracy.

The described physical magnitudes are chosen deliberately because they are in the basis of human senses: vision, hearing, touch, smell, and taste. Therefore, the presented optoelectronic devices designed on the unified principle can play the essential role for the artificial intelligence engineering.

REFERENCES

[1] Fedorinin, VN. RF *Patent*, 1999, 2, 157 513.
[2] Fedorinin, VN; Sokolov, VK. *Optics and Spectroscopy*, 1991, vol. 70, 1169-1175 (in Russian).
[3] Golay, MJE. *Rev. Sci. Instrum.*, 1947, vol. 18, 357-362.
[4] Robinson, LC. *Methods of Experimental Physics: Physical Principles of Far-Infrared Radiation;* Academic Press: New York and London, 1973, Vol.10, 166-173.
[5] Rogalski, A. *Infrared Detectors*; Electrocomponent Science Monographs; Gordon and Breach Science Publishers: Singapore, 2000, Vol. 10.
[6] Brenner, N. Rader C. *IEEE Acoustics, Speech & Signal Processing*, 1976, vol. 24, 264-266.
[7] Guo, H; Burrus, CS; Sitton, GA. *Proc. IEEE Conf. Acoust. Speech and Sig. Processing (ICASSP)* 1994, vol. 3, 445-448.
[8] Rokhlin, V; Tygert, M. *SIAM J. Sci. Computing*, 2006, *vol.* 27, 1903-1928.
[9] Hall, NA; Okandan, M; Degertekin, FL. *J. of Microelectromechanical Systems*, 2006, vol. 15, 770-776.
[10] Ko, WH; Guo, J; Xuesong, Ye; Zhang, R; Young, DJ; Megerian, CA. *IEEE International Symposium on Circuits and Systems, ISCAS*, 2008, 1812-1817.
[11] Lee, J; Kim, HJ; Lee, SQ; Lee, SK; Ko, SC; Park, KH. *Electronics Letters*, 2008, vol. 44, 576-577.
[12] Schock, SG; Tellier, A; Wulf, J; Sara J; Ericksen, M. *IEEE Journal of Oceanic Engineering*, 2001, vol. 26, 677-689.

[13] Vadus, JR. *Oceanic Engineering Society Newsletter* 2002, vol. 37, (3), available: http://www.ieee.org/organizations/society/oes/html/ summer 02/index.html
[14] Pinto MA. *ibid.*
[15] Campanella, S; Phyfe, D. *IEEE Transactions on Audio and Electroacoustics,* 1968, vol. 16, 26-35.
[16] Frocht, MM. *Photoelasticity*; Wiley-Interscience: New York, 1948, Vol. 2.
[17] Doyle, JF. *Modern experimental stress analysis*; J. Wiley & Sons: Chichester, 2004, 158-170.
[18] Nye, JF. *Physical properties of crystals: their representation by tensors and matrices*; Oxford Science Publication; University Press: Oxford, 1985, 235-258.
[19] Burger, CP. In *Handbook on experimental mechanics*; Edited by Kobayashi A.S; PRENTICE-HALL, INC: Englewood Cliffs, NJ, 1987.
[20] Born, M; Wolf, E. *Principles of Optics*; Pergamon: Oxford, 1968.
[21] Visnapuu, ML; Duzdevich, D; Greene, EC. In *Modern Research and Educational Topics in Microscopy*; Microscopy Book Series N3 FORMATEX: Badajoz, Spain, 2007, Vol. 1, 297-308.
[22] Vasil'eva, LL; Kushkova, AS; Repinskii, SM; Fedorinin, VN. *J. of Analytical Chemistry.*, 2000, vol. 55, 764-769 (in Russian).
[23] Seliger, H; Hinz, M; Happ, E. *Curr. Pharm. Biotechnol.* 2003, vol. 4(6), 379-395.
[24] Bilitewski, U. *Methods Mol. Biol.,* 2009, vol. 509, 1-14.
[25] Liu, J; Cao, Z; Lu, Y. *Chem. Rev.,* 2009, vol. 109(5) 1948-1998.
[26] Cosnier, S; Mailley, P. *Analyst.,* 2008, vol. 133(8), 984-991.
[27] Vercoutere, W; Akeson, M. *Curr. Opin. Chem. Biol.* 2002, vol. 6(6), 816-822.
[28] Dixon, MC. *J. Biomol. Tech.,* 2008, vol. 19(3), 151-158.
[29] Tombelli, S; Minunni, M; Mascini, M. *Methods,* 2005, vol. 37(1), 48-56.
[30] Willets, KA; Van Duyne, RP. *Annu. Rev. Phys. Chem.,* 2007, vol. 58, 267-297.
[31] Murphy, CJ; Gole, AM; Hunyadi, SE; Stone, JW; Sisco, PN; Alkilany, A; Kinard, BE; Hankins, P. *Chem. Commun. (Camb).* 2008, vol. 7(5), 544-557.
[32] Vladár, AE; Radi, Z; Postek, MT; Joy, DC. *Scanning,* 2006, vol. 28(3), 133-141.

[33] Fotiadis, D; Scheuring, S; Müller, SA; Engel, A; Müller, DJ. *Micron*, 2002, vol. 33(4), 385-397.
[34] Liu, ZC; Zhang, X; He, NY; Lu, ZH; Chen, ZC. *Colloids Surf. B. Biointerfaces* 2009, vol, 71(2), 238-242.
[35] Tang, J; He, Q; Chen, H; He, N. *J. Nanosci Nanotechnol.*, 2005, 5(8), 1225-1229.
[36] Drummond TG, Hill, MG; Barton, JK. *Nat. Biotechnol.*, 2003, vol. 21(10), 1192-1199.
[37] Moeller, R; Fritzsche, W. *IEE Proc. Nanobiotechnol.*, 2005, vol. 152(1), 47-51.
[38] Levina, A; Pyshnaya, I; Repkova, M; Rar, V; Zarytova, V. *Biotechnol. J.*, 2007, vol. 2, 879-885.
[39] Diamandis, EP; Hassapoglidou, S; Bean, CC. *J. Clin. Lab. Anal.*, 1993, vol. 7(3), 174-179.

INDEX

A

absorption, 7, 41, 57
adsorption, 37, 40, 42
AIDS, 45
aluminium, 7
amplitude, 11, 12, 13, 14, 16, 18, 21, 22, 23, 28, 35, 36, 38, 41, 43, 52, 58
artificial intelligence, 58
attachment, 47

B

background, 38
base pair, 45
basilar membrane, 18
biotin, 51, 52, 55
birefringence, 27
bonds, 40
boundary surface, 34
brain, 18

C

character, 17
chemical properties, 38
chemical reactions, vii, 2, 43, 57
circularly polarized light, 3
cochlea, 18
color, iv
communication, 1, 16, 18
complementary DNA, 47, 56
composition, 17, 40
compounds, 41
conductivity, 7
configuration, 38
consumption, 4, 6, 57
copyright, iv
cost, 16, 18
covering, 7, 56
crime, 56
crystalline, 10
crystals, 60

D

damages, iv
decomposition, 22
deformation, 7, 11, 12, 27, 28, 29
derivatives, 49
desorption, 37, 40, 42, 47
detection, 16, 29, 41, 43, 45, 47, 53, 56, 57
deviation, 43
diffusion, 40, 42, 43
diffusion process, 40, 43
displacement, vii, 2, 10, 13, 14, 15, 16, 28, 57

DNA, viii, 33, 45, 46, 47, 48, 49, 50, 51, 52, 53, 56, 57
double helix, 45, 46
drying, 51, 53

E

electric current, 1
electric field, 3, 4, 18, 35
electrodes, 47
electromagnetic, vii, viii, 1, 3, 4, 35, 57
electromagnetic field, 57
electromagnetic waves, 35
electron, 27, 38, 48
emission, 10
encephalitis, 53
energy consumption, 43
engineering, 17, 31, 58
equality, 8
equilibrium, 22, 42
equipment, 40
evaporation, 48
excitation, 18, 20
exposure, 16, 42, 43

F

FFT, 17
fiber, 1, 16
fiber optics, 1
films, 3, 28, 41, 47, 48, 53, 56, 57
flexibility, 13, 14
fluorescence, 38
fragments, 45, 47, 51
frequencies, vii, 17, 18, 20, 24, 58

G

gas sensors, 43
germanium, 27
grazing, 34

H

hair, 18, 57
hair cells, 18
height, 7
hepatitis, 45
homogeneity, 53, 56
hybridization, 45, 46, 47, 51, 52, 53, 54, 55, 56

I

illumination, 38
image, 7, 10, 12, 13, 16, 21
immobilization, 51
incidence, 33, 34, 35, 36, 40
information processing, 1
inhomogeneity, 38, 51
initial state, 52
integration, 22
intellect, 2
interaction process, 41
interface, 1, 33, 34, 36, 38, 40, 41, 47, 48, 51
interference, 10, 12, 13, 47

K

kinetics, vii, 2, 43, 57

L

lasers, 1
LED, vii, viii, 10, 29, 38
ligand, 47
light beam, vii, 33, 57
light emitting diode, 1
liquid chromatography, 40
liquids, 57

Index

M

magnetic field, 1
mammal, 18
manufacturing, 16
matrix, vii, 7, 8, 9, 10, 14, 16, 21, 22, 23, 57
media, 37
membranes, 10
metallurgy, 31
microscopy, 38, 47
mixing, 20
modulus, 28
molecular biology, 45
molecules, 40, 41, 42, 46, 47, 57

N

nanometer, 14
nanometers, 7, 10, 14, 37
nucleic acid, viii, 45, 56, 57
nucleotides, 45

O

opportunities, 1
optical properties, 47
optimization, 56
optoelectronics, 1
oscillation, 17, 21, 22, 23, 47, 58
oscillations, 17, 18, 20, 21, 22, 25

P

parallel, 18, 33
parameter, 1
PCR, 47
permission, iv
photodetectors, 3, 29
photoelastic effect, 27
photoelectron spectroscopy, 47
photons, 1
polarizability, 40
polarization, vii, viii, 1, 3, 4, 5, 10, 11, 34, 47, 53, 57
polyimide, 7
polymerase, 47
polymerase chain reaction, 47
probe, 29, 30, 45, 46, 50, 51, 52, 53, 54, 55
propagation, 34
properties, 60

Q

quality control, 40
quartz, 10, 27, 28, 47

R

radiation, vii, viii, 2, 7, 10, 11, 12, 16, 57, 59
radius, 10
reactions, 51, 52
reading, 16, 18, 52
receptors, 47
recognition, 17, 18, 25
recommendations, iv
reflection, viii, 4, 5, 21, 33, 34, 35, 40
refraction index, 10
refractive index, 2, 27, 28, 33, 34, 36, 37, 38, 40, 41, 47, 52, 57
reliability, 1, 51
relief, 29
requirements, 53
resolution, vii, 2, 16, 17, 18, 58
respect, 33
response time, 16
rights, iv

S

saturation, 41, 43
sediment, 51
selectivity, 45

semiconductor, 1, 16, 43
senses, 58
sensitivity, vii, 6, 13, 14, 18, 20, 28, 41, 43, 47, 51, 56, 57
sensors, vii, 2, 17, 37, 47, 53, 56, 57
signals, 17, 24, 51, 53, 56
signal-to-noise ratio, 14
silicon, 27, 53
Singapore, 59
SiO_2, 7, 48, 53, 55, 56
SiO_2 surface, 55
solid state, 1
sorption, 10, 47, 48
space, 3
Spain, 60
speech, 17, 25
sponge, 52
storage, 1
substrates, 51, 52, 53, 54, 55
sulfur, 41
sulfur dioxide, 41
surface layer, 47

T

temperature, 8, 13, 14, 16, 57
tension, 18
titanium, 53
total internal reflection, 33, 34, 37, 38, 40, 41, 48, 51
transformation, 21
transformations, 16, 23, 28, 38, 41, 52

transmission, 1, 5, 35
trapezium, 18
tuberculosis, 45
tungsten, 18, 19

U

uniform, 10

V

vacuum, 1, 53
vector, 3, 4, 35
vision, 2, 58

W

wavelengths, 7
wires, viii

X

xenon, 7
X-ray, 47

Z

zirconium, 53